High-Performance Double Skin Façade Buildings

This book provides a comprehensive theoretical platform for the use and construction of double skin façade projects. The DSF concept has been used mostly in European buildings; however, its success in other climates should be addressed. Increasing numbers of buildings are featuring double skin façade technology in the US; however, still relatively few have been studied for their performance in operation.

This book gives architects a practical guide to analyze and evaluate the actual performance of double skin façade buildings in different climatic contexts. It is important for high-performance buildings to have tools to evaluate a design's predicted performance to achieve specific sustainable goals. To determine that the application of DSF in different climates will provide better thermal comfort, building simulation tools analyze various thermal comfort parameters through studies of the façade and compare them with the actual building's performance data. The book takes the reader on an on-site tour of eight DSF buildings around the US. Interviews with the buildings' architects and engineers, owners, and users offer additional perspectives and insights into the construction and performance of these developments in building design.

This will provide architects with a comprehensive understanding of the challenges and opportunities in integrating double skin façades into their projects.

Mona Azarbayjani is the Graduate Program Director and an Associate Professor at the University of North Carolina Charlotte, School of Architecture, where she is currently teaching comprehensive studios, core building technology courses, energy modeling, and interactive envelope seminars. Her research is focused on energy analytics, high-performance and healthy buildings, and user-centered design with new technologies.

High-Performance Double Skin Façade Buildings

Climatic-Based Exploration

Mona Azarbayjani

Routledge
Taylor & Francis Group
NEW YORK AND LONDON

Cover image: Mona Azarbayjani

First published 2023
by Routledge
605 Third Avenue, New York, NY 10158

and by Routledge
4 Park Square, Milton Park, Abingdon, Oxon, OX14 4RN

Routledge is an imprint of the Taylor & Francis Group, an informa business

Library of Congress Cataloging-in-Publication Data
Names: Azarbayjani, Mona, author.
Title: High-performance double skin façade buildings : climatic-based exploration / Mona Azarbayjani.
Description: New York, NY : Routledge, 2022. | Includes bibliographical references and index.
Identifiers: LCCN 2022003159 (print) | LCCN 2022003160 (ebook) | ISBN 9781138857346 (hardback) | ISBN 9781138857353 (paperback) | ISBN 9781315718750 (ebook)
Subjects: LCSH: Exterior insulation and finish systems. | Facades—Design and construction. | Buildings—Environmental engineering. | Buildings—Thermal properties. | Sustainable buildings.
Classification: LCC TH2238.7 .A93 2022 (print) | LCC TH2238.7 (ebook)
DDC 693.8/32—dc23/eng/20220408
LC record available at https://lccn.loc.gov/2022003159
LC ebook record available at https://lccn.loc.gov/2022003160

ISBN: 978-1-138-85734-6 (hbk)
ISBN: 978-1-138-85735-3 (pbk)
ISBN: 978-1-315-71875-0 (ebk)

DOI: 10.4324/9781315718750

Typeset in Adobe Caslon Pro
by Apex CoVantage, LLC

Contents

List of Figures

CHAPTER 1

Introduction

There is a great deal of interest in using double skin façade (DSF) strategies in new and retrofitted buildings in Europe as it provides many possibilities for energy efficiency and, at the same time, creates better thermal comfort. However, the lack of performance assessment results in less popularity of implementing DSF in the US.

This book will trace the evolution of the double skin façade systems from early instances in the US (the Occidental Chemical Building 1982) to today's advanced configurations. The primary purpose is to provide an overview of various theoretical backgrounds regarding different aspects of building façade systems (building science and aesthetics) and their impact on building performance. First, a general definition of a double skin façade system is presented. Then a brief history, which describes the façade's evolution from early concepts of DSF to the present, is highlighted. The following section explains different DSF typologies and classifications regarding airflow driving force, facade divisions, airflow modes, and the presentation of previous work. The final section overviews seven different case studies' thermal performance considerations and envelope performances. The energy performance results help designers make better and more informed decisions for the type of façades that they are incorporating. The last chapter discusses the future of the building envelopes and describes examples of dynamic façades.

Detailed analysis of case studies in different climates on existing structures was completed in the US, with an evaluation of their performance compared to a curtain wall system being demonstrated. The list of buildings studied is as follows:

1. Cambridge Public Library, new addition – Boston, MA.
2. John E. Jaqua Academic Center for Student Athletes – Eugene, OR.

DOI: 10.4324/9781315718750-1

3. The Richard J. Klarchek Information Commons, Loyola University – Chicago, IL.
4. The Medical School's Biomedical Science Research Building (BSRB) – Ann Arbor, MI.
5. Seattle Justice Center – Seattle, WA.
6. Eli and Edythe Broad CIRM Center for Regenerative Medicine and Stem Cell Research at USC.
7. Occidental Chemical – Niagara Falls, NY.
8. Manitoba Hydro Place – Winnipeg, Ontario.

The architects of each project and the owner have been interviewed, and all the data have been collected. To understand the performance of each project, these projects were modeled and analyzed by an energy modeling tool, and their actual performance was compared with their projected performance. The DSF concept has been used primarily on European buildings; however, its success in other climates of the US needs to be addressed. This book gives architects a practical guide to analyze and evaluate the actual performance of double skin façade buildings in a different climatic context. High-performance buildings need to have tools to assess a design's predicted performance in order to achieve specific sustainable goals. Although the number of buildings featuring double skin façade technology has increased, there are still relatively few buildings which have had their performance studied in operation.

In this book, the application of DSF in different climates will provide a better understanding of the performance of the façade through using building simulation tools. It analyzes the energy performance and thermal comfort parameters through studies of the façade and compares them with the as-built/actual building's performance data.

It presents three significant contributions to the evaluation of the performance of DSF buildings:

- A methodology for assessing the energy and comfort performance of DSF buildings through energy analysis and airflow modeling was developed based on 3-dimensional analysis.
- Models to simulate the specific DSF typology and couple the envelope level results to a building simulation program.
- A framework for comparing and evaluating the façade solution for different climatic contexts.

The book takes the reader on an on-site tour of the seven DSF buildings around North America. The author also interviews the architects and speaks with engineers, owners, and users to get additional perspectives. By evaluating and understanding the buildings, architects discover the challenges and opportunities in integrating double skin façades. Finally, it discusses a vision for the future of building envelopes and addresses a number of questions related to the façade's technology. Emerging technologies present opportunities to fundamentally shift current expectations of building envelope functionality to building component systems. The envelope can be transformed in response to occupants' needs while meeting the energy demands of buildings. Recent envelope systems include dynamic façades, providing the opportunity to consider individual comfort needs rather than collective comfort in the design process.

Overview

Reduction of energy consumption has emerged as a major challenge in the US. The search for better approaches in improving thermal comfort conditions and the energy efficiency of buildings is intensifying. Currently, low-energy building design features include lighting and controls, ventilation systems, and an improved building envelope. Lighting energy can be effectively reduced in several ways: through the use of natural daylighting, high-efficiency fixtures and controls (such as occupancy sensors that turn lights off when there is no movement), and photosensors that reduce light output as needed to maintain a minimum level. These technologies, combined with architectural details like light shelves, highly efficient glazing facades, and external shading, can increase natural daylight while reducing energy consumption associated with artificial light.

The use of glass and transparent materials is not a new technique, and in fact, has become an attractive envelope option primarily in high-rise buildings. Large windows and all-glass exteriors are popular due to the natural light and views to the outside they provide, but this attractive design element has pitfalls, in particular, solar heat gain.

Solar heat gain that occurs when sunlight strikes the windows and passes through them creates discomfort and increases the cooling energy consumption, since air conditioning is needed to sustain a comfortable environment indoors.

The building façade plays an essential role in achieving high-performative buildings. Energy-consuming systems required for providing fresh air to meet indoor air quality requirements can be reduced or eliminated using passive or hybrid technologies. Hybrid ventilation, or the use of natural and mechanical systems to cool and ventilate buildings, offers opportunities to take advantage of external conditions but requires a backup system to maintain the indoor environment when these conditions are not adequate.

Climate influences the feasibility and usage period of natural ventilation as a means of cooling. Buildings in temperate

Figure 1.1 Interior Thermal Comfort

climates can use natural ventilation for most of the cooling season, which is usually from May through October. In climates with a broader temperature range, including hot summers, passive cooling is still possible, but greater attention to detail and design must be paid. The location of the building affects performance as well.

The majority of energy in built environments is used in heating, cooling, and ventilating occupied spaces. The situation is even more critical for commercial buildings.

In office buildings, the use of natural ventilation techniques can be a potential design-strategy component. Heating and cooling loads could be decreased by incorporating energy-efficient technologies, control schemes, and improved building-envelope design. Designers and engineers can continue improving the building envelope and façade treatments to reduce heat loss and solar gain through the windows, decreasing heating and cooling requirements and minimizing differences between indoor air and surface temperatures that may cause occupants' discomfort. The increased use of thermal mass to temper indoor air temperature has become more widely used in commercial building design. In hot summer continental climates like Chicago, the energy consumption required to cool and ventilate a building can be reduced by incorporating natural ventilation into the building design during the shoulder seasons, though it is not often done. In most climates suitable for natural ventilation, only a handful of buildings have this type of system.

The challenge for designers is to improve building performance while providing a more comfortable and healthier place for users. Fortunately, there are numerous methods and techniques that can be employed to achieve these goals.

Heating and cooling loads can be decreased by incorporating energy-efficient technologies, control schemes, and improved building-envelope design. Designers and engineers can continue improving the building envelope and

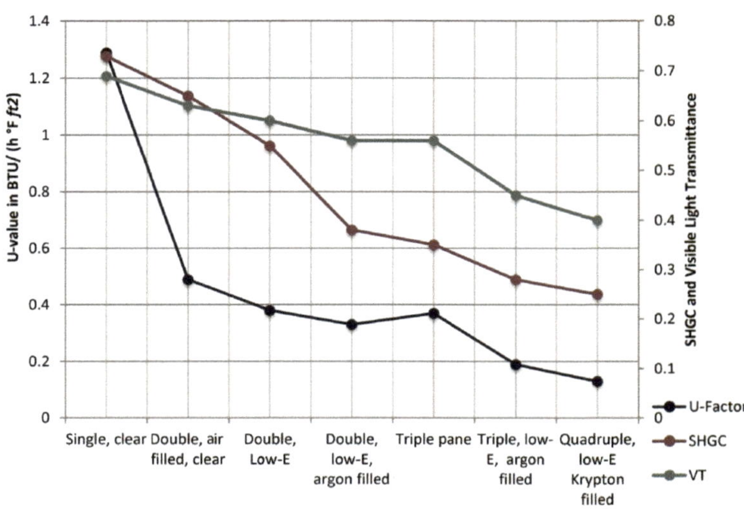

Figure 1.2 SHGC and U-Value Comparisons of Different Façade Systems

façade treatments to reduce heat loss and solar gain through the windows, decreasing heating and cooling requirements and minimizing differences between indoor air and surface temperatures that may cause occupants discomfort.

Developing new façade technology is necessary for more environmentally friendly and energy-conscious building design. While a great deal of interest exists in learning how to integrate double skin façade systems in our buildings, there is little knowledge or demonstration of how this system performs in the US.

This book attempts to analyze the DSF configurations in different climatic contexts of the US in terms of energy and thermal comfort. The main objective is to explore the design and construction practices that promote energy efficiency, cost-effectiveness, and comfort in built projects around the US.

History

European architectural trends mainly drive the concept of DSF, and it dates back to when many central European houses utilized box-type windows to increase thermal insulation [1]. In 1849, Jean-Baptiste Jobard, then director of the Industrial Museum in Brussels, described an early version of a mechanically ventilated multiple skin façade. He mentioned

Figure 1.3 Stieff Factory and the Birth of the Curtain Wall System

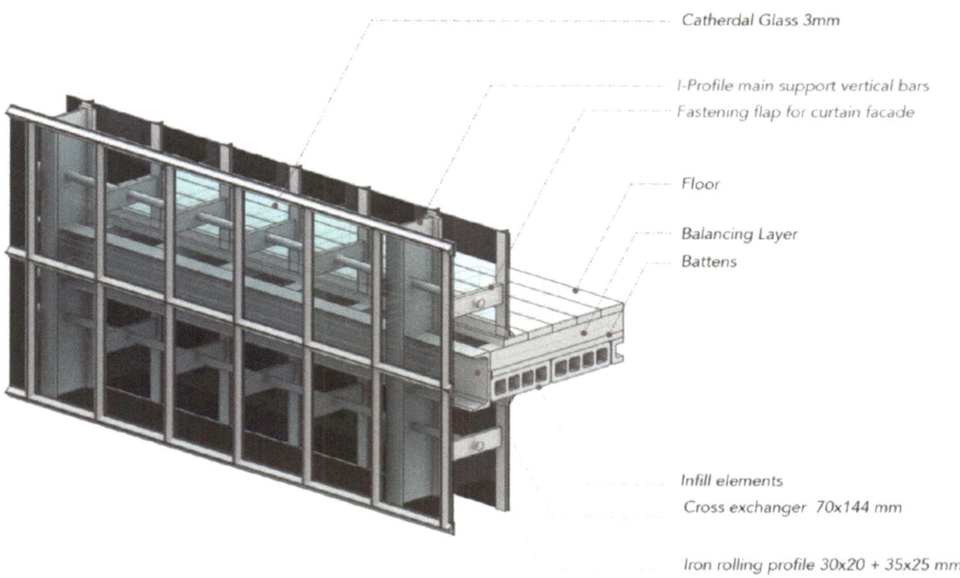

Figure 1.4 Double Skin Façade

that in winter, hot air could be circulated between two glazings, while in summer, cold air could be used [2]. The Stieff Factory was built in response to the increasing international demand for felt toys. Teddy bears and toys were the reason behind the construction dimensions of 100ft long, 40ft wide, and 13ft high, with the exterior materials being a double-glazed façade with a flat roof. This structure provides natural daylighting while considering the cold climate and strong wind velocity in the area.

After the construction of these factory buildings, the double skin façade developed and traveled to Russia, where it was used in the communal housing building of Narkomfin (1928). Later, Le Corbusier started using two membranes with a gap in between, while integrating mechanical systems.

There was little progress on the double skin façade construction until the early 1980s. During this period, green buildings gained attention, and multi-layered glass façades, as a solution, influenced architectural design decisions.

A DSF was assumed to be energy efficient and environmentally friendly, providing a 40 to 60 percent reduction in energy consumption, external noise reduction, and natural ventilation, even in skyscrapers (Oesterle et al., 2001). Aside from an aesthetic consideration, such concepts are often chosen – particularly in high-rise buildings – for shading devices, protection from high wind pressure, and the possibility of natural ventilation due to lower pressure in the façade cavity.

Double skin façade is more common in Europe than in the US due to higher energy prices in Europe and stricter energy codes; however, after early 2000, double skin façades became integrated into higher-end office buildings. The context for using double skin needs to be significantly studied, as high humidity and temperature will not be an appropriate environment for this concept.

Single vs. Double Skin Façade

A single skin façade in typical buildings consists of different functional layers, including cladding, structure, insulation, etc., which separates the building interior from the exterior. In the nineteenth century, the emergence of curtain walls, which did not have a load-bearing function, made it possible to form an envelope made entirely of glass. Nevertheless, the issues related to thermal comfort conditions and the low thermal performance of those lightweight single skin envelopes were supported by the advent of mechanical systems.

The energy crisis of 1973 necessitated the designer exploiting renewable energy sources for cooling, heating, lighting, and ventilating a building. Those issues resulted in the development of new products in the glass industry to protect against unwanted heat gain in the summer and optimize the use of daylight. Examples of such products include glass with different coatings. At that time, the architects attempted to incorporate

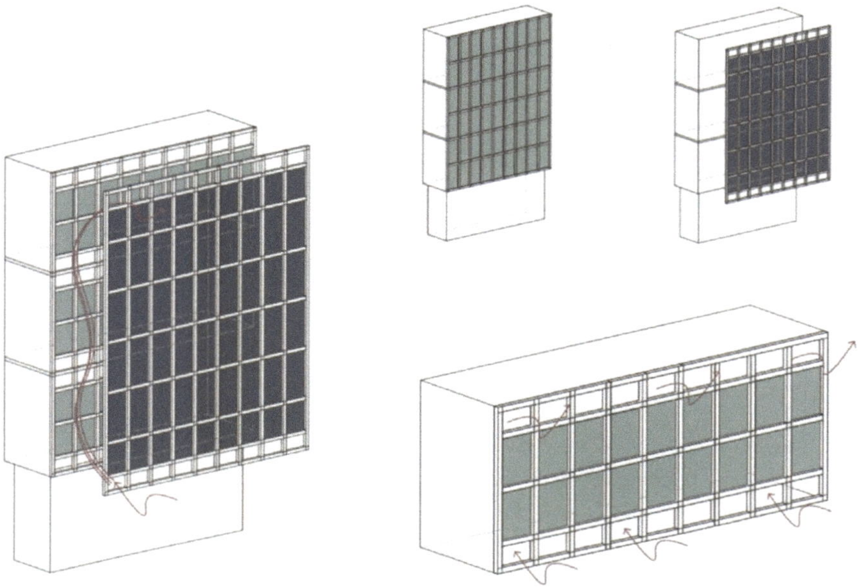

Figure 1.5 Single Skin vs. Double Skin Façade

the use of natural ventilation and solar energy, which resulted in the emergence of new concepts in façade systems such as double skin façades.

A double skin façade consists of a multilayered façade with an external and internal layer that contains a gap between the two layers used for controlled ventilation and solar protection. Ideally, double skin façades are designed as an integral part of the HVAC system; therefore, they play a significant role in energy and thermal comfort efficiency.

The use of multilayered skins deploys building insulation against thermal variation and external noise. As illustrated in Figure 1.5, a DSF could provide outdoor air through operable windows and preheat or pre-cool the air introduced to the interior space by the HVAC system. It can integrate mechanical ventilation with natural ventilation during the mild seasons. Psychologically speaking, it can reconnect inhabitants with the outdoors and provide a healthy environment while using a substantially smaller mechanical system.

In this introduction, a definition of double façades was introduced. A brief history of DSF is also presented. The next chapter will outline the existing typologies based on the origin, the driving force of the airflow, and divisions of the façade. It also allows for a description of how a DSF might operate. The energy and thermal performance assessment of different DSF examples will be studied in Chapter 6.

References

1 Oesterle, E., Leib, R.D., Lutz, G., Heusler, B. (2001). Double skin facades: integrated planning: building physics, construction, aerophysics, air-conditioning, economic viability. Prestel.

2 Saelens, D. (2002). Energy Performance Assessments of Single Storey Multiple-Skin Facades. Ph.D. thesis, Laboratory for Building Physics, Department of Civil Engineering, Catholic University of Leuven, Belgium. Retrieved from: http://envelopes.cdi.harvard.edu/envelopes/content/resources/pdf/case_studies/PhD_Dirk_Saelens.pdf

Bibliography

Gerber, D. J., Pantazis, E., & Wang, A. (2017). A holistic approach for performance-based architecture: Design exploring geometry, user, and environmental agencies in façades. *Automation in Construction*, 76, 45–58.

Pérez-Lombard, L., Ortiz, J., & Pout, C. (2008). A review on buildings energy consumption information. *Energy and Buildings*, *40*(3), 394–398. https://doi.org/10.1016/j.enbuild.2007.03.007.

Saelens, D., & Hens, H. (2001). Experimental evaluation of airflow in naturally ventilated active envelopes. *Journal of Thermal Envelope and Building Science*, *25*(2), 101–127.

Saelens, D., Carmeliet, J., & Hens, H. (2001). Modeling of Air and Heat Transport in Active Envelopes, Proceedings of ICBEST 2001, International Conference on Building Envelope Systems and technologies, Ottawa, Canada, pp. 243–247.

Saelens, D., Carmeliet, J., & Hens, H. (2003). Energy performance assessment of multiple skin facades. *International Journal of HVAC&R Research* *9*(2): 167–186. Retrieved from: http://www.bwk.kuleuven.ac.be/bwf/pdf_artikels/I_J_HVACR_DS_2003.pdf

CHAPTER 2

Basics

Types of Construction – Box, Shaft, Corridor, Multi-Story Types

Classification of DSF

Three characteristics identify types of double skin façades. These are based on geometric characteristics and also the ventilation mode and type.

As presented in Figure 2.1, these criteria are:

1. Type of ventilation
2. Partitioning of the façade
3. The modes of cavity ventilation

The differences between the pressure created by the stack effect and the wind effect are the two main driving forces of natural ventilation. Natural ventilation drives air through space by taking advantage of the pressure differences caused by these two factors, due to internal temperature differences or a combination of both.

Hybrid ventilation is a controlled compromise between natural and mechanical ventilation. In hybrid ventilation, natural ventilation is generally used; mechanical ventilation is only triggered when the driving forces of natural ventilation become inadequate and make it no longer possible to achieve the desired performances. Note: The stack effect (or chimney effect) is a phenomenon related to the rising of hot air. Applied to a DSF, the concept of stack effect expresses the fact that warm air is lighter than cold air. As the cavity is generally hotter than the outside, the air tends to escape at the top.

Based on the ventilation mode, the DSF was categorized into five

DOI: 10.4324/9781315718750-2

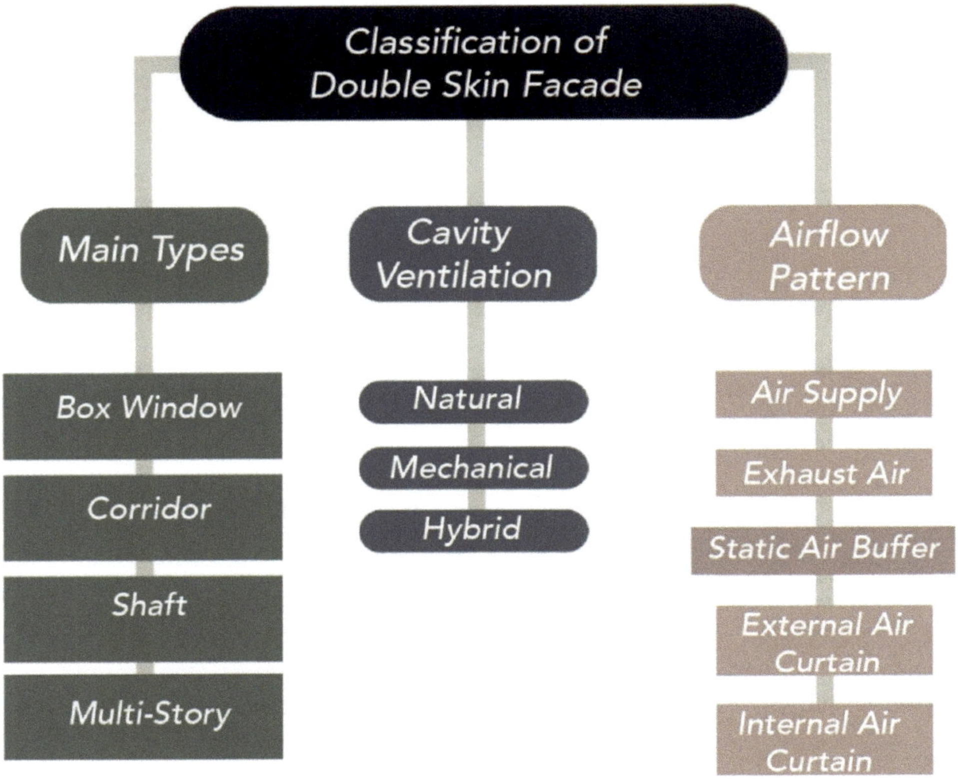

Figure 2.1 Classification of DSF

groups. Not all façades can adopt all the ventilation modes described here; a single ventilation mode characterizes a façade.

However, a façade can adopt several ventilation modes at different moments, depending on whether or not specific components integrated into the façade permit it (i.e., operable openings). One must distinguish between the following five main ventilation modes:

- Outdoor air curtain: In this ventilation mode, outside air introduced into the cavity is immediately turned back outside. The cavity ventilation thereby forms an air curtain enveloping the external façade.
- Indoor air curtain: The air comes from the inside of the room and is returned to the same place or via the ventilation system. The cavity ventilation forms an air curtain enveloping the indoor façade.
- Air supply: The ventilation of the façade is created with outdoor air. The air then travels to the interior room or into the ventilation system. This makes it possible to supply the building with air.
- Air exhaust: Air inside the room is evacuated towards the outside. The

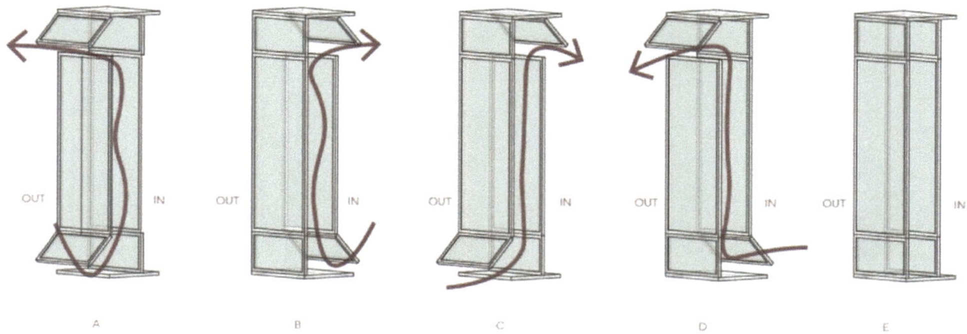

Figure 2.2 DSF Classifications Based on Ventilation Mode

ventilation of the façade thus makes it possible to evacuate air from the building.

- Buffer zone: This ventilation mode is distinctive, as each of the double façade skins is made airtight. The cavity thus forms a buffer between the inside and the outside, with no possible cavity ventilation.

Partitioning the Façade

Once the decision on the type of ventilation has been made, the geometry of the cavity and how the air flows will need to be determined. Based on the 'façade's geometry (width openings, cavity height, width, etc.), double skin façades are categorized into the following groups: box window, shaft window, corridor façade, and multi-story [1].

If the DSF extends over the entire height and width of the building, the term, façade, is appropriate. Three main categories can be defined if the façade is divided into smaller units. The expression "shaft façade" is adopted if the partitioning consists of vertical ducts. The term "corridor façade" is usually employed when the façade is horizontally partitioned. If the façade is both horizontally and vertically subdivided, the DSF is called a window or box. The term "windows" can be used for systems in which the windows act as DSFs. The term "box" is more appropriate for entirely transparent envelopes with horizontal and vertical partitioning.

Box Type

The façade is horizontally and vertically subdivided in this type, with entirely transparent envelopes. Horizontal and vertical partitioning divides the façade into smaller and independent boxes. A box window is the oldest form of DSF and consists of a frame and inward-opening casements.

The cavity between two façade layers is divided horizontally and vertically along construction axes and floors, respectively. This type of window is common in areas with high external sound levels with particular requirements relating to sound insulation between adjoining rooms. This form is the only type that provides the function of a DSF with a conventional means of opening [1].

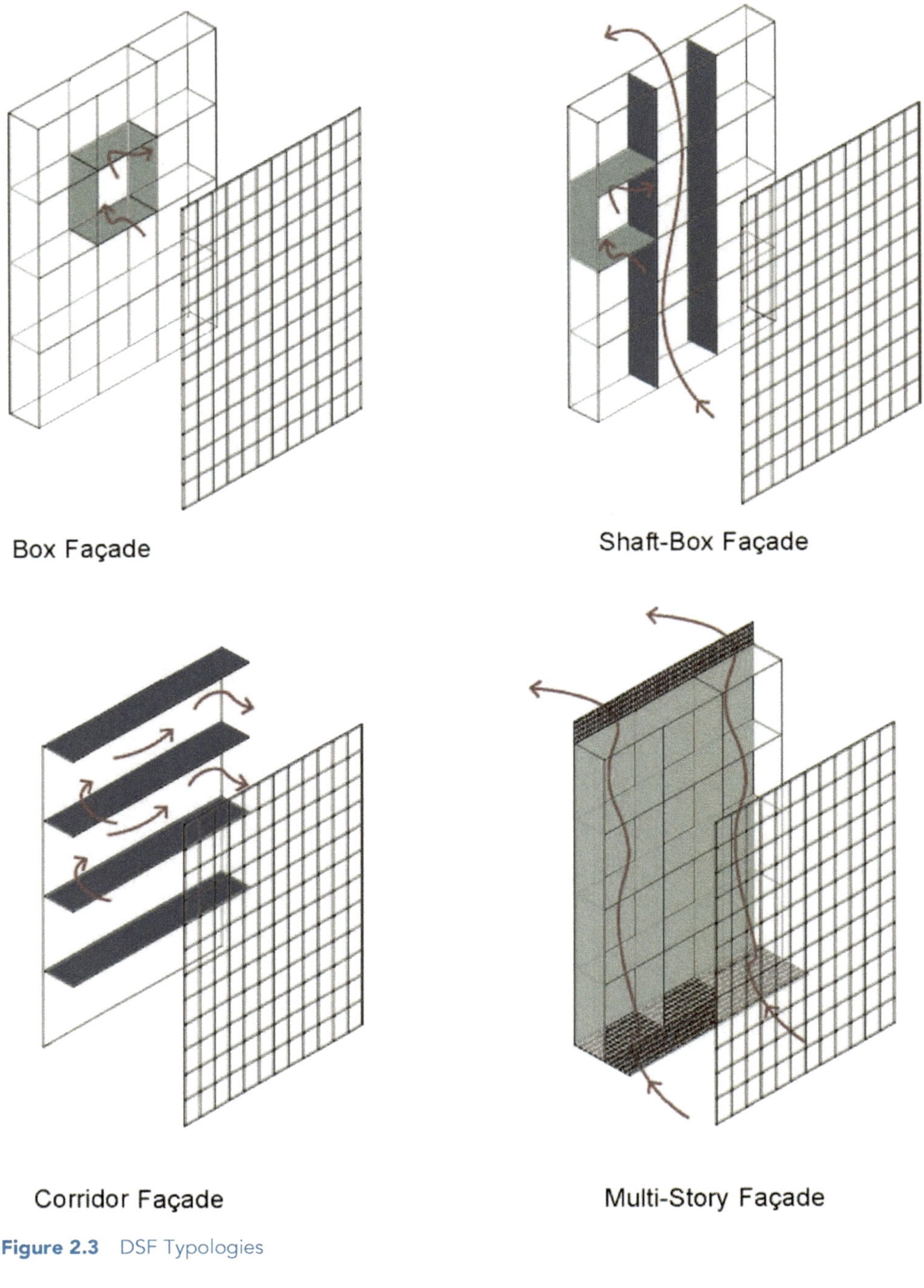

Figure 2.3 DSF Typologies

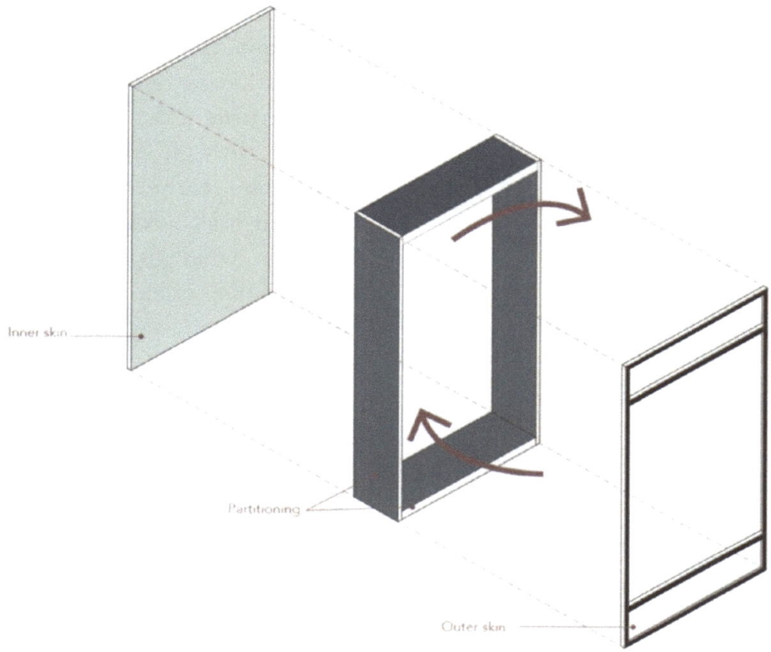

Figure 2.4 Box Type Façade

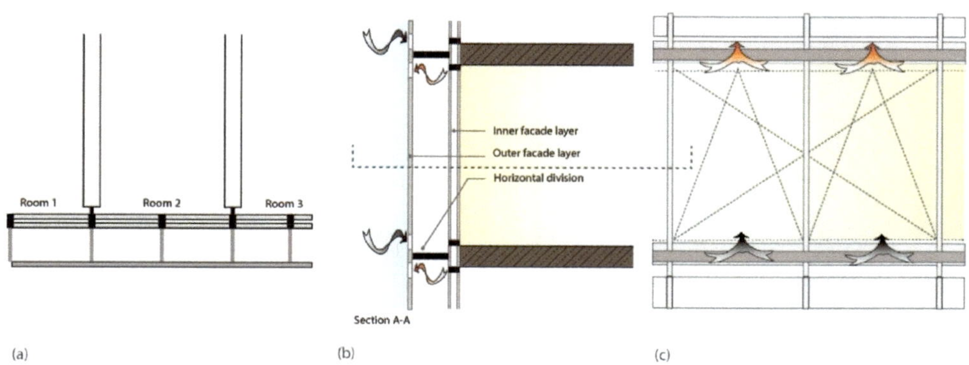

Figure 2.5 Box Type Window a) Plan b) Section c) Elevation

Both inner and outer skins can be fixed or have operable windows to incorporate natural ventilation or otherwise create a buffer zone during the cold months.

Each box can be prefabricated and be installed on the job site. However, this configuration is quick to construct and is suited for areas with more strict fire and acoustics requirements. The only setback for this type is not providing enough thermal buoyancy due to the geometry.

Shaft Type

In this case, a set of box-window elements are placed in the façade with continuous vertical shafts that go along several stories to create a stack effect. On every level, the vertical shafts are linked with the adjoining box windows by an opening.

The stack effect draws the air from the box window into the vertical shafts. The air can also be sucked out mechanically.

This type of DSF is suited for low- to mid-rise buildings since the height of stacks is limited [1].

Figure 2.6 shows the plan, elevation, section, and ventilation diagram in the shaft type. In order to use the shaft type in a high-rise, we must divide the shaft by horizontal partitions every 7–9 floors instead of having one continuous shaft.

The arrows indicate the route of the airstream through the box windows and the common shaft.

The air inlet is located at the bottom of the box windows, and the exhaust is located at the higher level on the sides of the box and connected to the shaft. The shaft extracts the air via thermal buoyancy that rises and creates a negative pressure at the bottom of the shaft that exhausts the air from the box windows. The inner skin of the shaft type can be operable or fixed. However, in order to take advantage of the buoyancy effect, the chimney's skin

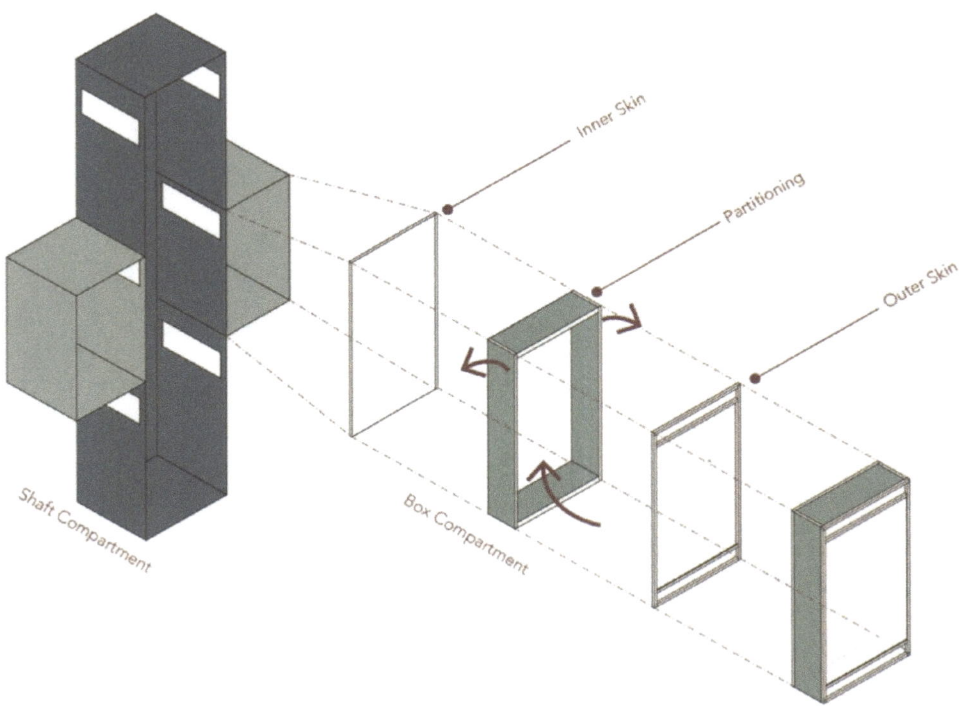

Figure 2.6 Shaft Type Double Skin Façade

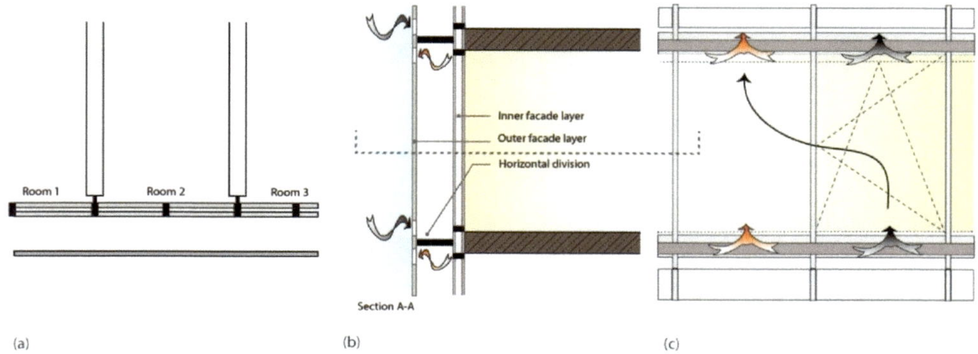

Figure 2.7 Shaft Type Double Skin Façade a) Plan b) Section c) Elevation

should not be operable (though accessible for maintenance reasons).

Heat can be recovered at the top of the chimney, which is one of the additional benefits of the shaft type.

The disadvantage of using this shaft type would be the possibility of fire spreading and noise transmission through the chimney as it rises through a number of floors.

Corridor Type

When necessary, divisions occur horizontally along the corridor for fire protection or ventilation reasons. The intake and extract openings are situated near the floor and the ceiling. They are usually staggered to prevent extracted air on the floor from entering the space on the floor immediately above [1]. In the corridor façade,

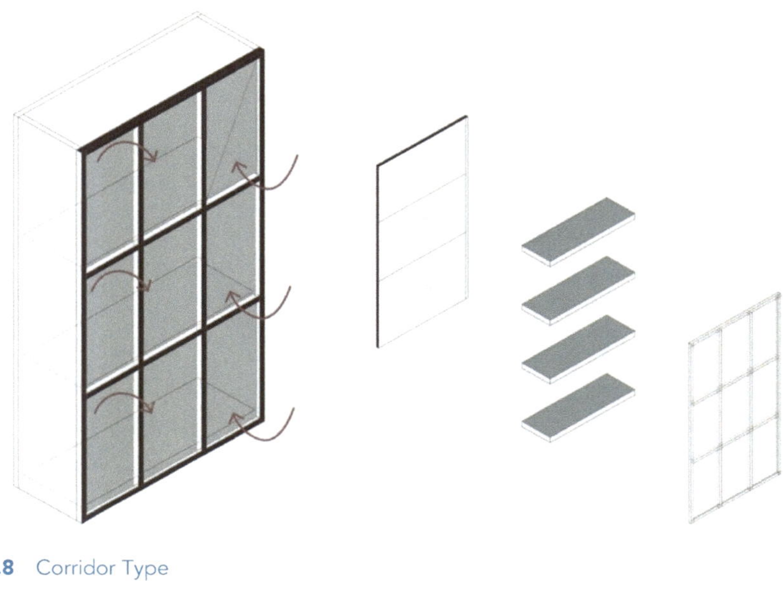

Figure 2.8 Corridor Type

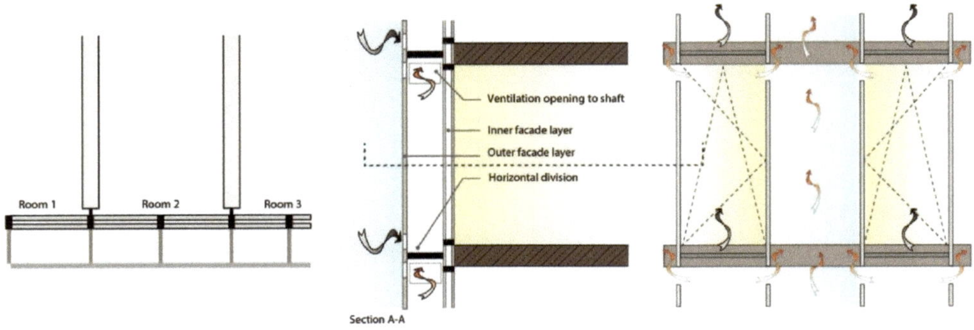

Figure 2.9 Corridor Type Double Skin Façade a) Plan b) Section c) Elevation

thermal buoyancy is not very strong. Since the cavity is divided horizontally at each floor, the chance of spreading noise and fire between floors is minimized. Natural ventilation can be incorporated through the operability of the two inner and outer skins. The corridor also provides an exterior space in front of each room. A plan, section, and elevation of a corridor façade are illustrated in Figure below.

As shown, the intermediate space is not divided at regular intervals along its horizontal length. Air flows diagonally to prevent extracted air from the lower story from being sucked in with the air supply of the above floor. The section of the corridor façade shows the separate circulation for each level.

As shown in the plan, the air cavity can create noise transmission between the spaces on the same floor.

Multi-Story Type

This is the most common type of DSF in the US. In this DSF case, the cavity is

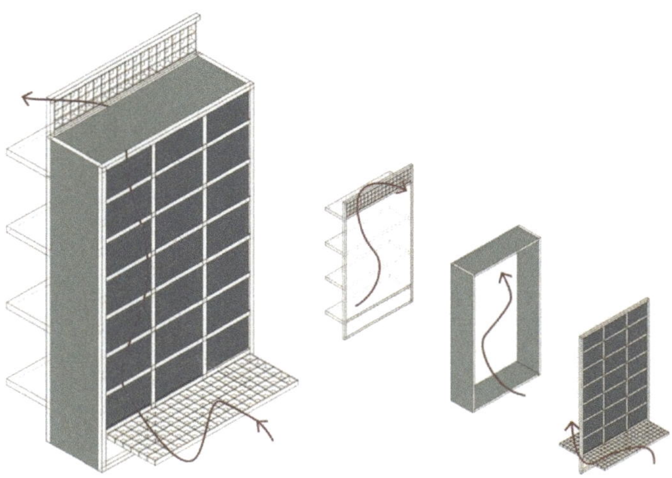

Figure 2.10 Multi-Story Double Skin Façade

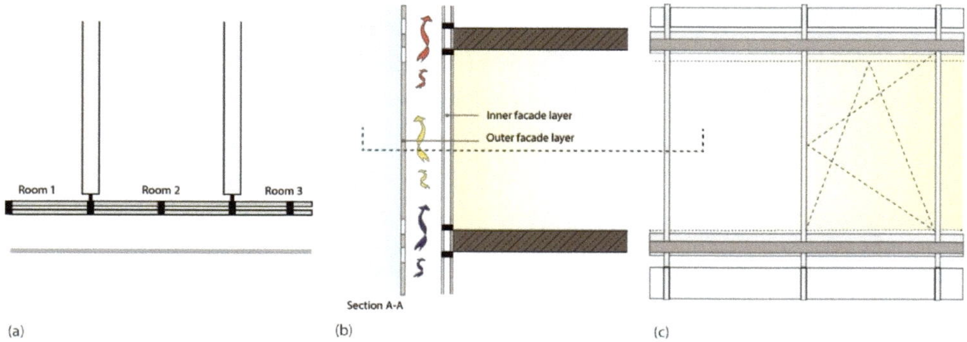

Figure 2.11 Multi-Story Type Double Skin Façade a) Plan b) Section c) Elevation

adjoined vertically and horizontally by a number of rooms.

Ventilation occurs via large openings near the ground floor and roof. The room behind the DSF should be ventilated mechanically [1].

For winter conditions, the cavity can be closed at the top and bottom to take advantage of the greenhouse effect created within the cavity. During the summer, the cavity is kept open to exploit cooling buoyancy. The warmer air within this space is lighter and rises upward, thus causing a state of excessive pressure at the top. In contrast, the heated air provides less pressure at the bottom of the opening, which forces it into the intermediate space. This results in less comfort in the upper floors if the heat is not totally ejected.

References

1 Oesterle, E., Leib, R.D., Lutz, G., Heusler, B. (2001). *Double skin facades: Integrated planning: Building physics, construction, aerophysics, air-conditioning, economic viability*. Prestel.

Bibliography

Crespo, A.M.L. History of the Double Skin Facades. Retrieved from the web: http://envelopes.cdi.harvard.edu/envelopes/content/resources/PDF/doubleskins.pdf

Loncour, X., et al. (2005) Best Practice for Double Skin Façades, WP 1 Report "State of the Art" Reporting Period: 1.1.2005–31.12.2005, Editor: Wolfgang Streicher, Institute of Thermal Engineering, Graz University of Technology (WP1 Leader).

Roth, K., Lawrence, T., Brodrick, J. (2007). Double-skin facades. *ASHRAE Journal*, *49* (10), pp. 70–73.

CHAPTER 3

Façade Tectonics

Façade Construction

This use of glass as a predominant element of the building façade is not new, and the concept of all glass façades goes back to the Crystal Palace (1851). Structural glass façades technology clearly has its roots in the great iron and glass conservatories of the 19th century.

The term "curtain wall system" refers to an architectural envelope set forward of the plane of structure and where there is no expression or interruption by the structural system. This type of system does not merely refer to hanging a sheet of glass in front of a building. Instead, it consists of metal panels with windows set within them.

A curtain wall is a type of vertical building façade system that supports no other forces except its own weight and the environmental forces which act upon it. This use of glass as a predominant element of the building façade exploded in the 20th century, fueled by Modernism, especially post-war Modernism. After some initial stunning architectural innovations like the Bauhaus Building in Dessau by Gropius in 1926, different Mies van der Rohe buildings in 1954, and the Lever House by SOM in 1952, this ultimately led to all-glass façade building towers in any climatic context. Later on, closer to 2000, it led to double skin façade development and component façades. To understand the construction of a double skin façade, first, the structure and evolution of the curtain wall system will be described and followed by the development of the double skin façade system.

Post-and-Beam Façade

It consists of story-high posts linked by horizontal beams. The space between posts and beams can perform as cladding providing lighting and ventilation. The post and beam provide stability against wind forces and support the cladding and structural weight.

DOI: 10.4324/9781315718750-3

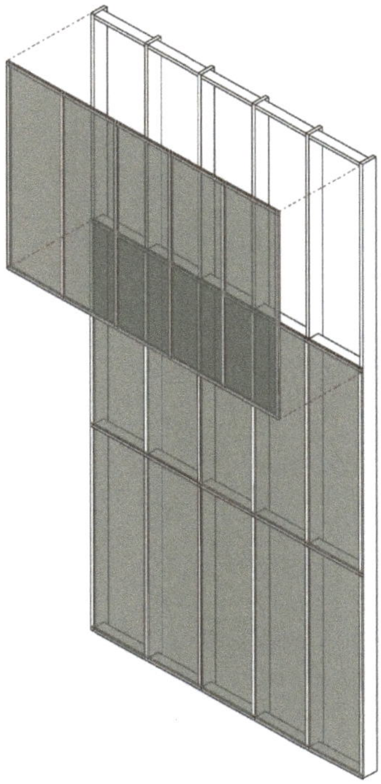

Figure 3.1 Post and Beam Façade

Element Façade

The curtain wall can be divided into stick or unit systems. Depending on the construction, the whole system can be prefabricated off-site and installed on-site.

In the stick system, the vertical mullions and horizontal transoms are installed on aluminum brackets which are anchored to the columns/slabs. Cut-to-size glass is fixed on the grid

The system is installed with pressure plates. (Shown in Figure 3.2).

The long pieces of aluminum – that is where the name stick came from – are inserted vertically and horizontally between vertical members to support and transfer the load of the glass back to the structure. Most of the erection of the stick system is done on the job site and involves the installation of long pieces

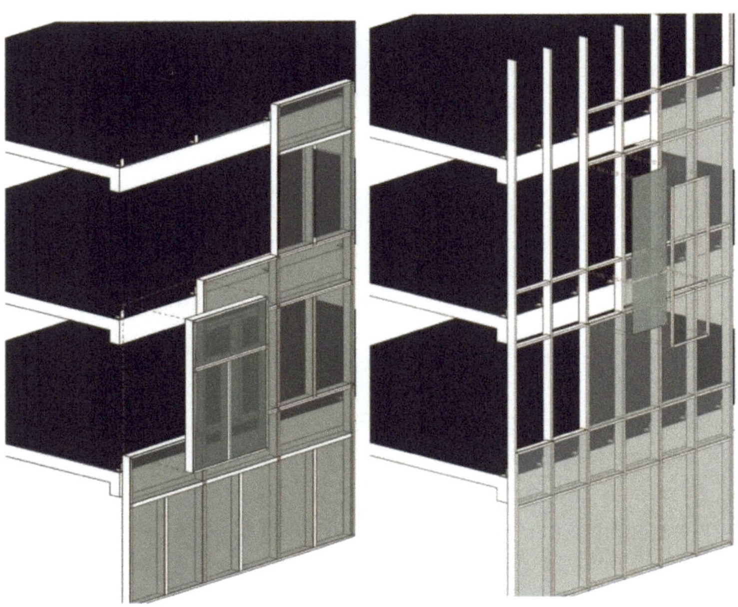

Figure 3.2 Unitized System (left), Stick System (right)

between floors vertically and between vertical members horizontally.

The unit system, the complete unit spanning the floor height entirely, is fabricated at the factory and installed on the brackets. This system involves the assembly of prefabricated panels hung from the structure of the building and is the quickest and easiest curtain wall system to assemble. Once on-site, the units can then be erected onto anchors connected to the building.

Double Façade

The next development is the double skin façade's system that accommodates internal functions such as daylighting

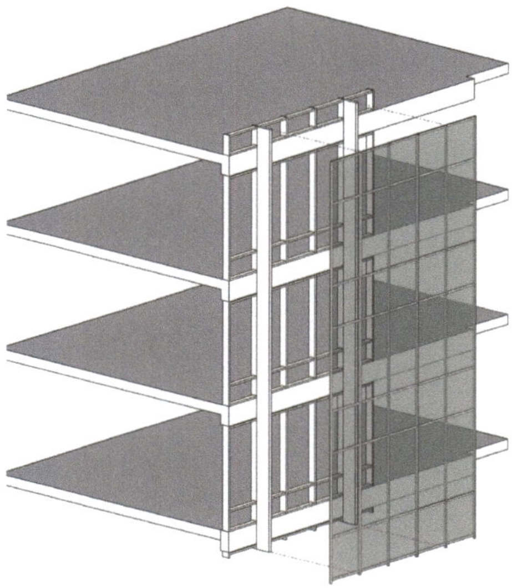

Figure 3.3 Double Skin Façade Structure

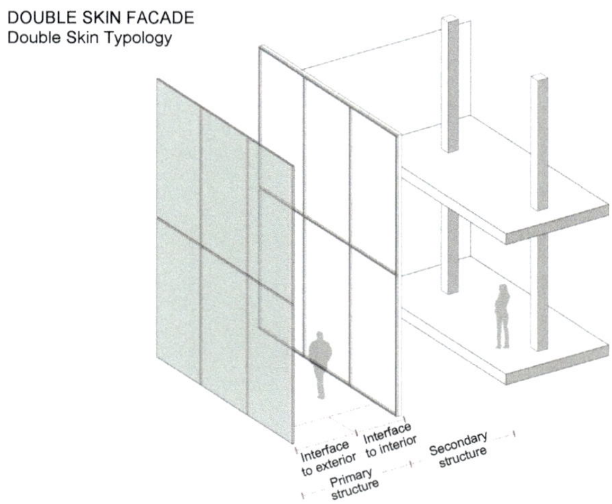

Figure 3.4 Double Skin Façade Layer

or ventilation. The only way to provide natural ventilation in high-rise buildings is through double skin façade systems. A double skin façade creates an extra layer of glazing in front of the façade system in order to provide natural ventilation and higher acoustic quality, as shown in Figure 3.4. Also, the shading devices can be integrated into this cavity and be protected from the exterior environment. In high-rise buildings, these advantages can make a double skin façade an economical solution as a façade system.

DSF Construction

The double skin façade construction comprises three layers: the primary structure of the building, the secondary structure that holds the façade and its elements, and the third layer are the façade elements.

The secondary structure of the façade transfers the load to the primary structure. The primary and secondary structures can be different (i.e., concrete and steel). However, they need to be coordinated if two other companies are going to erect them.

The whole post-and-beam structure can be a series of cables and rods, as shown in the image below. Like the post and beam construction, the structure can be without the posts or the beams or have both. The posts can be truss or cable rods.

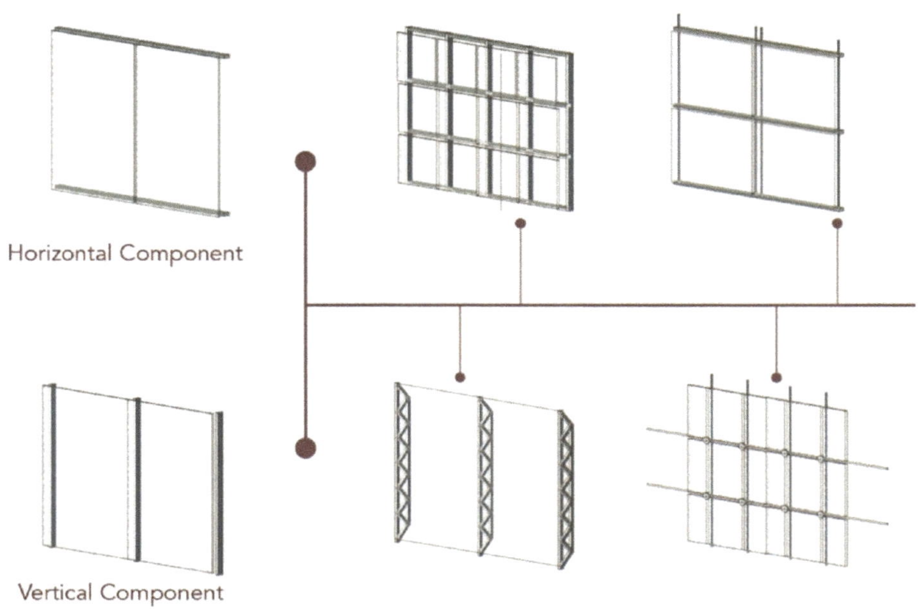

Figure 3.5 Types of Components for Façade Construction

Figure 3.6 Multi-Story Façade Construction

Double Skin Multi-Story, the Most Common Type in the US

The double skin multi-story façades are the most common type of DSF in the US. It consists of two layers of glass with a cavity in-between, usually comprised of one prefabricated unit system façade and the outer layer with its own structure that ties back to the primary structure of the building. Similar to the unit systems, the glass elements were pre-assembled to a great extent, and the mounting part of the façade was aligned before the units were installed.

The exterior skin consists of point-fixed, bolted, laminated glass, hung 3' off the surface, using a cable mullion system. The cables are anchored to a steel beam at the base. The steel cables are suspended from a cantilevered structure anchored to the roof, and T-shaped outriggers anchored to embeds on each floor extend to provide lateral support to the cables. These outriggers also support aluminum grating used for maintenance and cleaning purposes.

During installation, outriggers, base girders, and other boundary anchor steel components were installed after confirming the position of embeds on each floor. Once the prestressed vertical steel cables were hung and secured to the base, they were tensioned individually. After verifying the prestress forces on the cables, outriggers on individual floors were installed, followed by the installation of aluminum gratings. Cable clamps were then fixed to the cables, and bolted fittings were secured to the glass planes. The glass panes were lifted with a crane and placed in position, going from ground to top. The final step was the sealing of joints with structural silicone caulking.

Figure 3.7 Custom Aluminum Extrusions Support Double Skin Façade in Loyola Chicago, and Point Supported Glass

Figure 3.8 Aluminum Extrusions Support a Double Skin Façade

In another example of a double skin façade in Chicago, the interior skin utilizes a unique proprietary façade system. Custom aluminum extrusions span between floors without horizontal members, and glass panels are point supported through patch plate clamps. There is no flat supporting system in

Figure 3.9 Components of the Multi-Story Façade

this typology, and the edge of the glass is not connected to the mullion directly; instead, there is an extension-fitting that holds the glass in its place. Each module consists of three panels (vision, operable vent, spandrel).

The installation procedure followed a stick assembly with the mullions set first, followed by a dead load bracket at the bottom edge of the glass panel. The point-fixing clamps were then positioned and secured. Glass panels were set and secured with exterior clamp plates, and joints between panels were sealed using butt-glazed in situ silicone.

The outer skin of the double skin façade consists of a two-way cable net structure (a grid of 29 vertical and six horizontal pre-stressed stainless-steel cables) supporting 5' X 8' monolithic glass panels. The vertical cables are suspended from "hangman" outriggers at the top of the building (the hangmen-base is supported from and anchored back to the building roof structure with cable ties) and anchored to the base with a concealed spring assembly. Horizontal cable members are anchored through "shark fins" (plate anchors) on both sides and clamped to the vertical cables with a four-part cast stainless steel clamp. This cast assembly also doubles as support for the glass panels and is complemented by an exterior plate. A kicker strut ties each vertex to a building column.

The installation procedure began with the installation of anchors on all three sides. Horizontal cables were hung first, followed by suspension of the vertical cables, and tensioning of the cables followed the same order. Cable clamps were then installed after confirming the forces on cables. The glass panels were installed from top to bottom and finished with a silicone weather seal application.

Support Structure

The structure of the double skin façade falls into secondary and tertiary structures on how to hold the inner and the outer skins. The double skin façades in North America can be categorized into two main categories based on the structure that holds up the double skin system.

Cantilever/bracket

Structural frame

The two main categories each have two subcategories based on the attachment of the glazing to the primary structure: point connection and unitized system.

Cantilever/Bracket

A double skin structure is extended from the primary slab in this method. This extension can happen either by welding the cantilever profile to the intermediate floor slab using a cleat or bolting it to the slab concrete core. Using bolts for attachment benefits from enabling minor adjustments to the bracket in both directions by utilizing the slotted holes in the plate (Figure 3.10).

One can choose from two standard methods to attach glazing to the cantilever brackets. The first is using a point connection to the glass. In this method, the spider bracket is welded to the cantilevered bracket. Then a glass with drilled holes matching the point connection will be attached to the spider bracket. Around the hole, a sealant will be used in order to make sure the surface is leak-proof and also to provide a spacing buffer for glass deflection based on structural movement. The connection between the glazing to the structure does not happen at all the connection points. The number of connections is driven by the properties of the façade, such as area, loads, height, etc. All the other connections that are not connected directly are interconnected by the tension rods in vertical and horizontal directions. The tension rod goes through the spider node that holds the four edges of four different glass panes together. (Figure 3.14)

Figure 3.10 Cantilever/Bracket

Figure 3.11 Cantilever/Bracket

Figure 3.12 Structural Frame

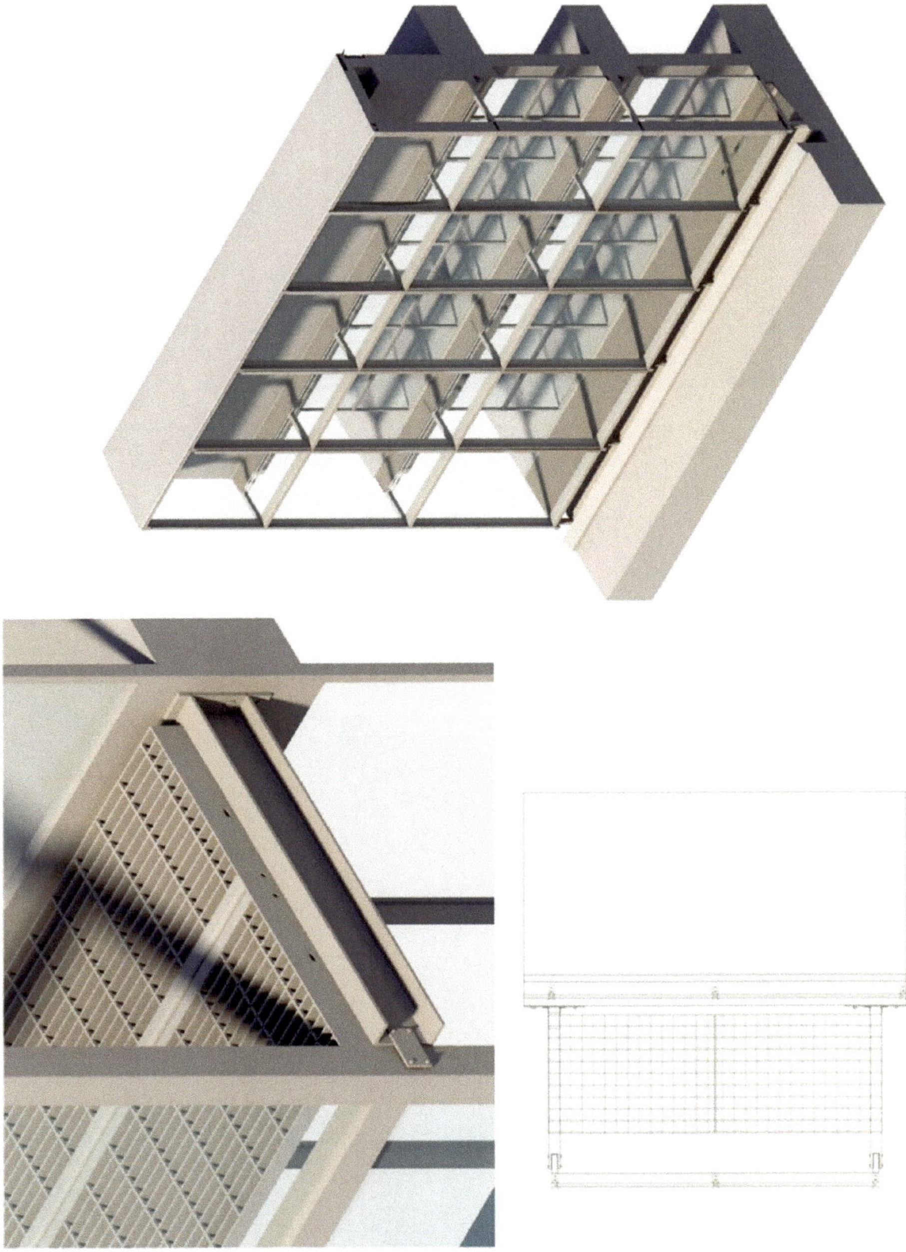

Figure 3.13 Cantilever Bracket with Unitized System, Detail (top left), Plan (bottom left), Cantilever Structure (right)

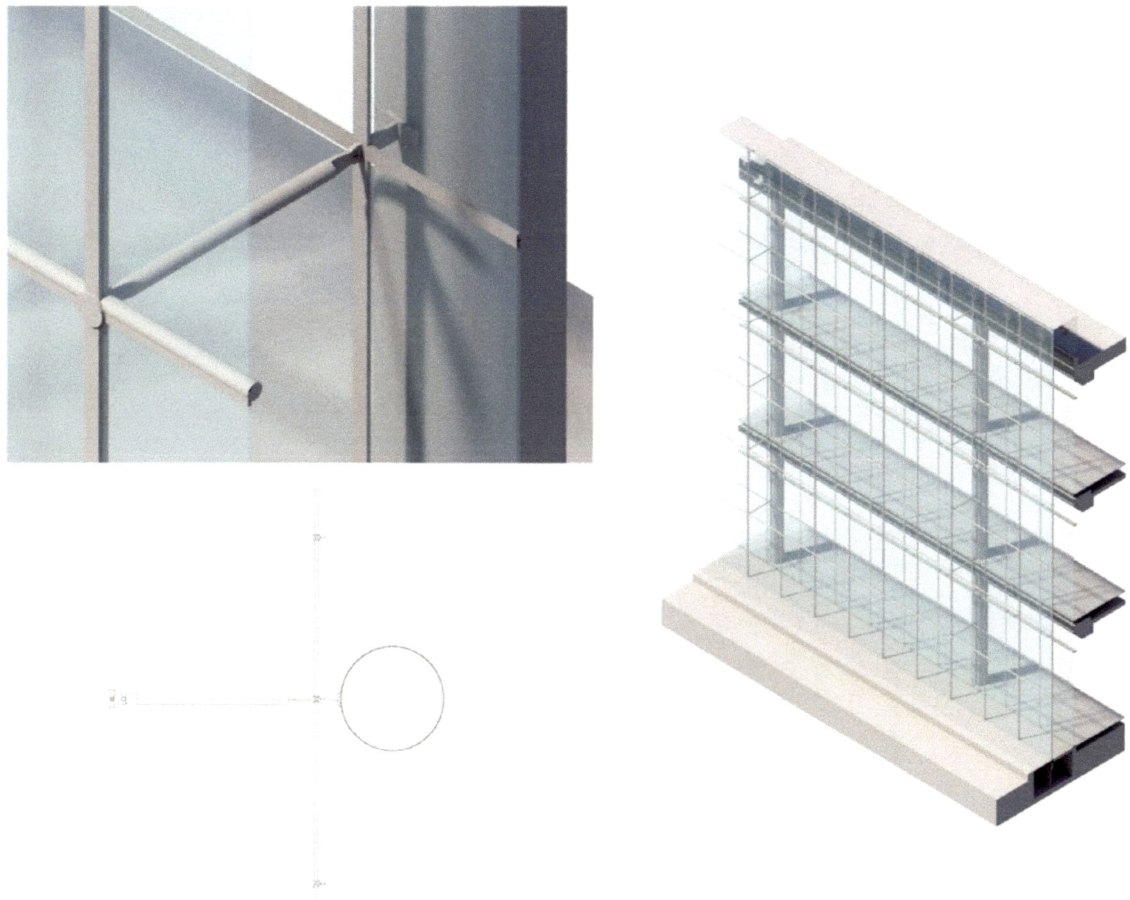

Figure 3.14 Cantilever Bracket with Point Connection, Detail (top left), Plan (bottom left), Bracket Structure (Point Connection) (right)

Another method of attaching the glazing is using unitized glass panes. The glass pane is placed in a rigid frame (plastic, timber, aluminum, steel). The gap between glass and aluminum is sealed using a sealing material. These unitized frames will be transferred to the job site and then will be attached to brackets. (Figure 3.13)

Structural Frame

In this system, the structure that holds the double skin is an array of rectilinear frames that are interconnected via horizontal members, which form a Truss. This structure is more rigid than the bracket structure. When a glass pane is attached to it, it is supported from all four sides; hence deflection of the glass pane is negligible. There are two ways of connecting the structural frame to the primary structure:

An extended cantilever is welded to the structural frame. Then the extension will be connected to the intermediate slab either by welding to the intermediate beam or using a cleat or bolts to connect the plate to the slab.

In order to attach the glazing to the frame, we can use two methods. One is to connect the glass directly to the structure using a point connection. In this method, brackets are welded directly to the structure; the glass with matching holes drilled into them will be attached to the brackets. The space between the glass and bracket is filled with a washer in order to seal it. A bolt will be screwed onto the bracket to hold the glass in its place on the outer side. (Figure 3.17)

The other attachment method is to use unified glass. In this method, glass which is framed by an aluminum tube will be contacted to the structure.

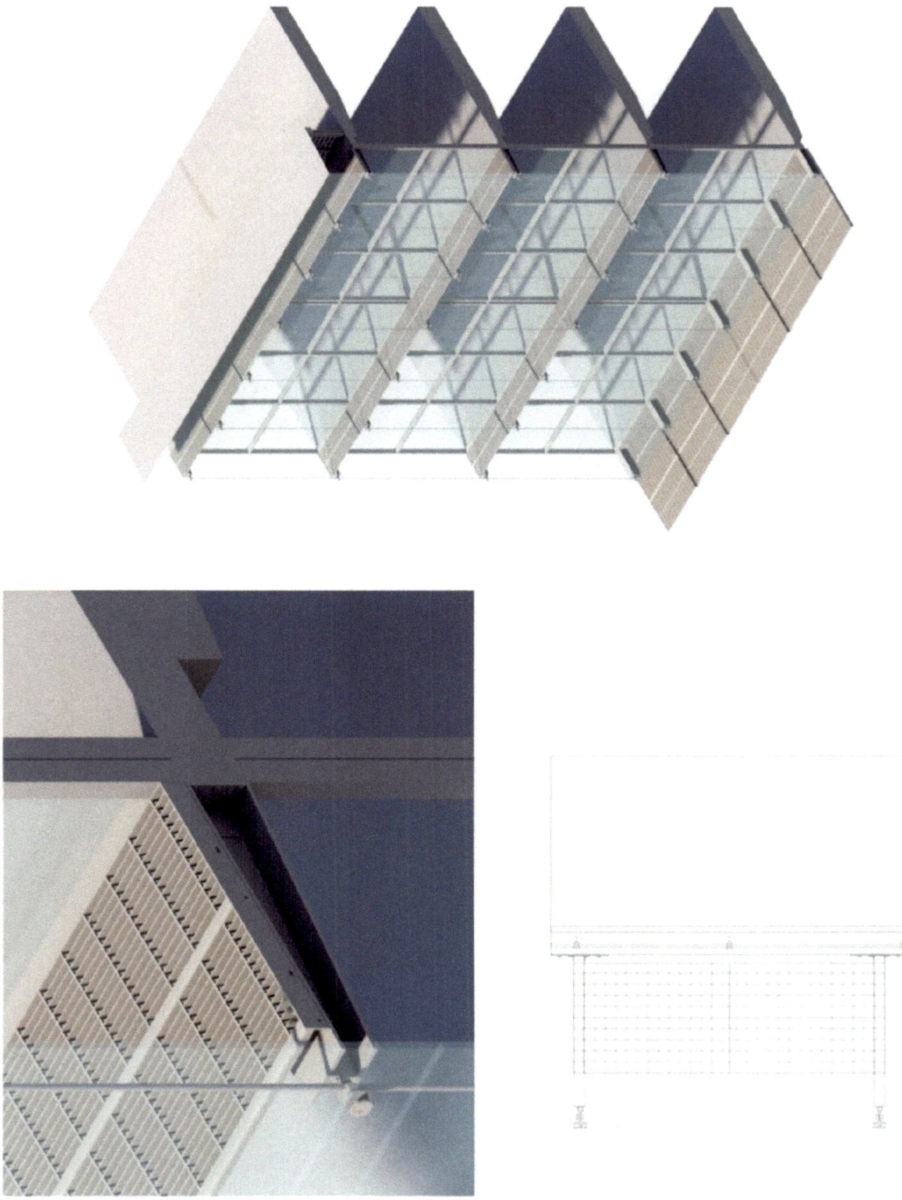

Figure 3.15 Structural Frame, Detail (top left), Plan (bottom left), Structural Frame (right) with Point Connection

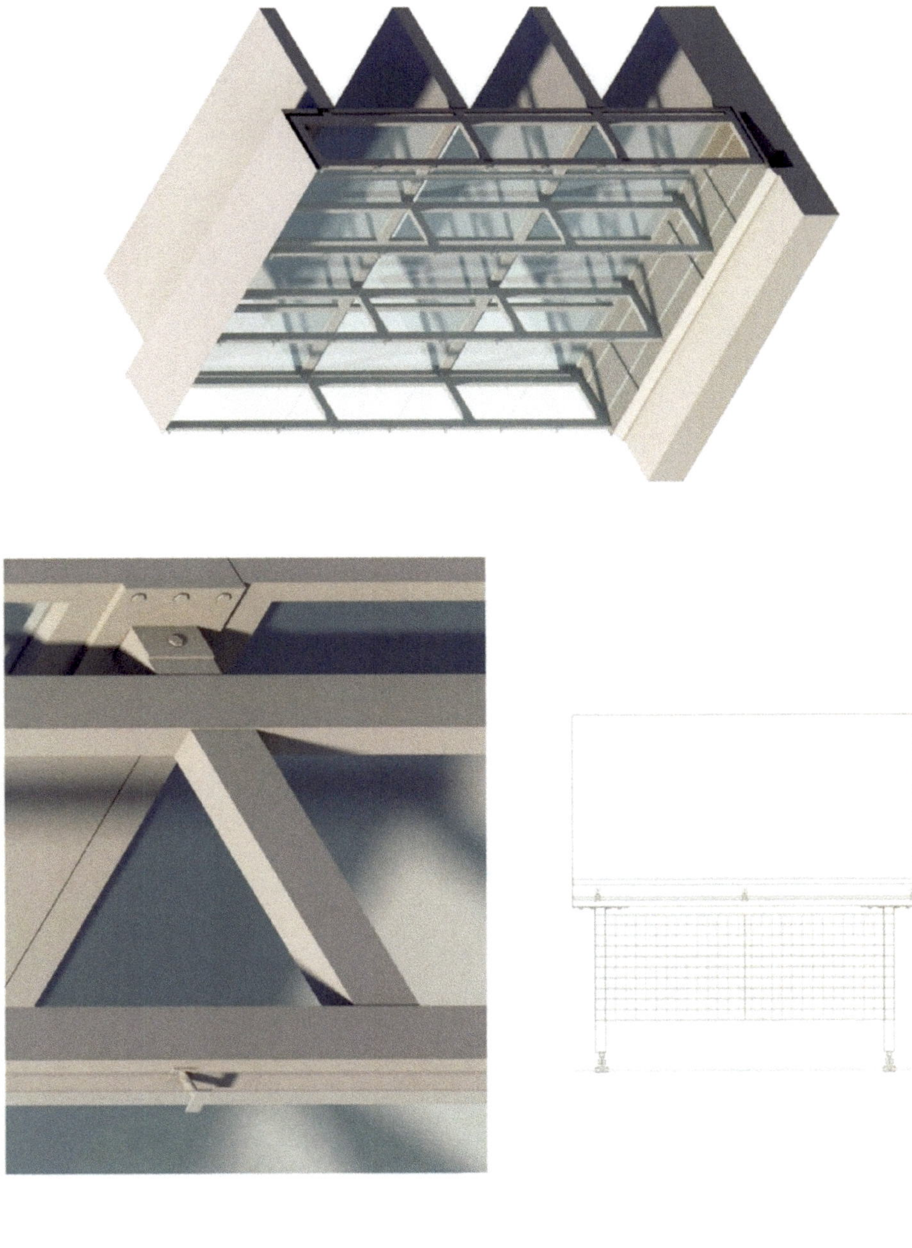

Figure 3.16 Detail (top left), Plan (bottom left), Structural Frame or Truss System (right)

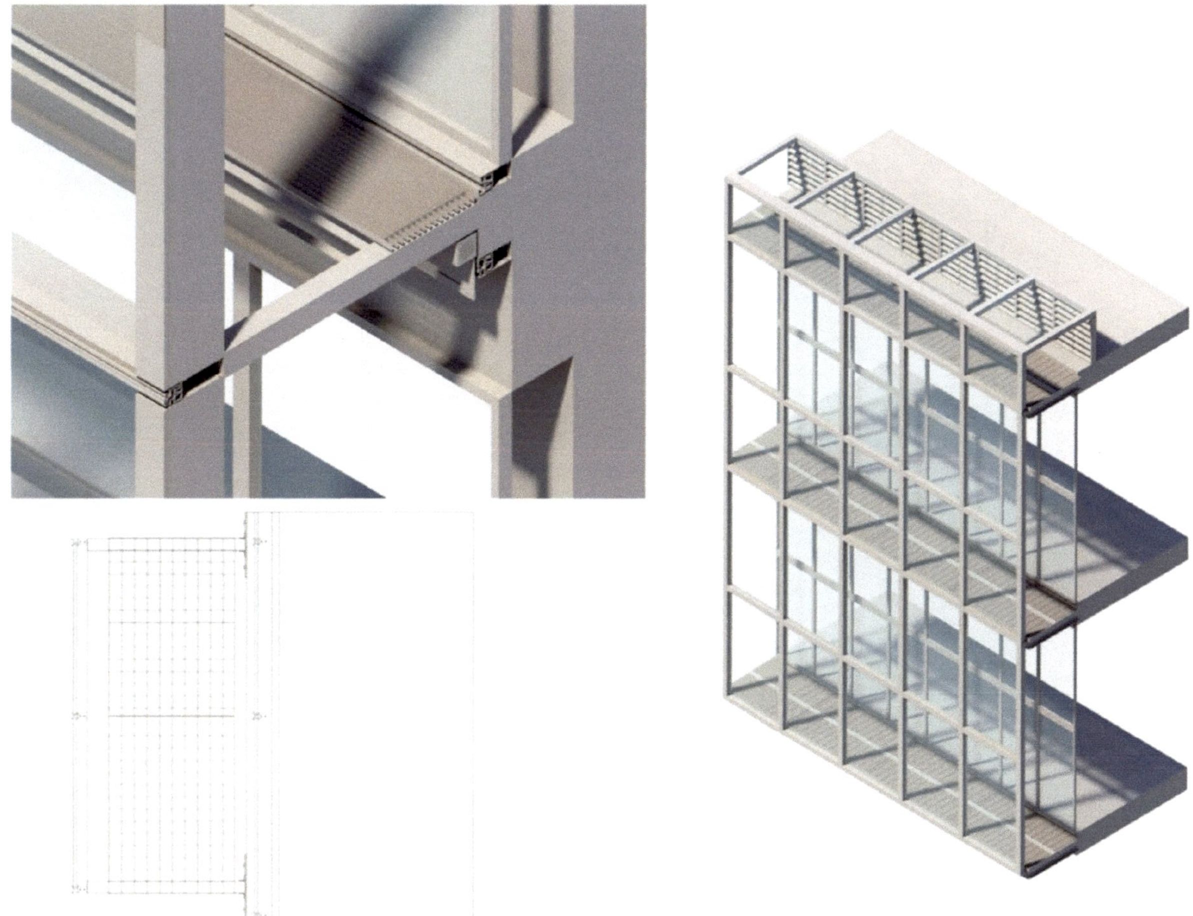

Figure 3.17 Detail (top left), Plan (bottom left), Structural Frame (right) with Point Connection

Bibliography

Kawai H., Oura, Y., Suzuki, M. & Nishimura, H. (2009). Field Measurement of Wind Pressure on a Double Skin with a Ventilator, in [5th European & African conference on wind engineering: Florence Italy, July 19th-23rd 2009, conference proceedings. - (Proceedings e report; 51)] [Firenze: Firenze University Press, 1534]. http://digital.casalini.it/2309039

Knaack, U. (2014). *Façades : Principles of construction* (Second and revised edition.). Birkhäuser.

Uuttu, S. (2001). Study of Current Structures in Double-Skin Facades. MSc thesis in Structural Engineering and Building Physics. Department of Civil and Environmental Engineering, Helsinki University of Technology (HUT), Finland. http://www.hut.fi/Units/Civil/Steel/SINI2.PDF

CHAPTER 4

Climatic Context and Design Solution

Climatic Condition

Due to varying climate regions, there are vast differences in energy consumption. Though the majority of energy consumption in building operation goes into environmental control systems and artificial lighting, the key to improving the energy performance also lies in the deployment of strategies to take those loads off the building's portfolio.

It must be noted that the design of high-performance buildings requires a deep understanding of climatic context potential and constraints.

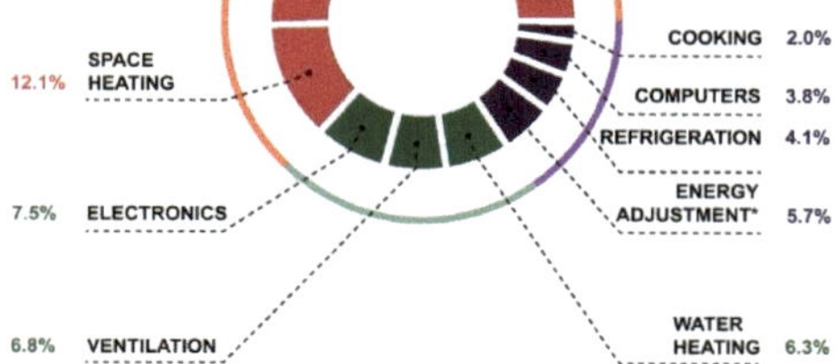

Figure 4.1 End-Use Energy Consumption in the US

Thermal Performance

Although the equivalent thermal transmission coefficient U-Value for a ventilated façade system will be worse to some extent (in comparison with a single skin façade), the results will enhance significantly if the intermediate space (cavity) is closed during the wintertime. During the heating period, the additional skin and the cavity of the DSF system provide enhanced insulation quality by increasing the R-Value of

DOI: 10.4324/9781315718750-4

the façade, because air also acts as insulation. The reduced airflow speed and increased temperature inside the cavity lower the heat transfer rate on the surface of the glass, which leads to a reduction of heat loss. This has the effect of maintaining higher temperatures on the inside part of the interior pane. Oesterle et al. (2001) describe the proportion of the opening area to improve thermal insulation.

During the summer, once radiation passes into the building, it is absorbed by the building fabric and re-radiated as long-wave infrared energy that does not pass back through the glass. As a result, the air in the cavity will be heated via convection. The hot airflow in the cavity can pass through the glazing outside and inside the space via Conduction. As the cavity warms up, the stack effect is improved respectively. A DSF system results in less heat transfer from outside to inside and less energy required to cool the space.

The U-value expresses the heat flux (in W/m^2K) through a building component (or a combination of components) with a temperature difference of one degree across several components under steady-state conditions. Figure 4.2 shows the heat transfer mechanism through single pane glass; the heat transfer mechanism through DSF is illustrated below. A low U-value indicates that the building component has a high thermal resistance and can be expressed in a case for a multiple-glassing unit by the following equation (Oesterle, 2001):

$$U = \frac{1}{\frac{1}{h_e} + \frac{1}{h_i} + \frac{1}{h_t}}$$

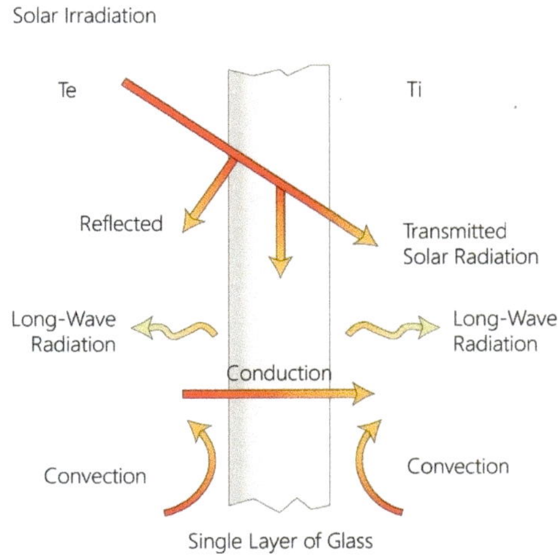

Figure 4.2 Heat Flux Through the Glass

Where:

$U = W/m^2K$

h_e = External heat transfer coefficient (W/m^2K)

h_i = Internal heat transfer coefficient (W/m^2K)

h_t = Conduction of multiple glazing units (W/m^2K)

The external heat source is solar radiation, which is initially reflected on the exterior glazed skin in this process. Depending on external conditions, the external heat transfer coefficient h_e is determined. The remaining radiation passes through the glass. The reflection with the cavity's inner glass and inner walls creates processes of convection and conduction, which determine the heat transfer coefficient inside the cavity h_i. The accumulated and remaining heat by radiation and conduction received by the room, determines the heat transfer coefficient h_t.

Different climate zones require different design strategies. The building envelope serves as the primary interface of the building, and along with the exterior environment, it plays an important role in energy consumption and occupants' comfort. Advanced technologies in façade construction enable buildings to respond to the environment effectively and take advantage of the site forces to develop passive strategies. These façade systems have the potential to capture, filter and redirect daylight, provide natural ventilation and take advantage of solar heat gain or block it. For each climate zone, all the possible passive strategies will be discussed. Façades/Windows account for >30 % of primary energy use in buildings, tall buildings, and buildings with High S/V Ratio account for more.

In choosing design strategies, the particular climate zone must be considered to minimize the impact or take advantage of contextual assets. For instance, heating-dominated climates benefit from the collection of solar radiation, passive heating, heat storage and retention (such as improved insulation to reduce heating demand), and the use of daylighting to reduce lighting demand. In cooling-dominated climates, protection from the sun and its direct solar radiation becomes more important.

Thermal Comfort

The thermal condition is one of the primary factors of Indoor Environmental Qualities (IEQ) that can cause Sick Building Syndrome (SBS) and influence occupants' health. Schedule-based and predefined environmental control is one of the main reasons for the current discomfort and dissatisfaction.

Thermal comfort models are an approach for quantification of human subjective evaluation and their thermal state that provides a common understanding of thermal satisfaction. Thermal comfort evaluation is a difficult task due to many influential variables such as air temperature, air velocity, relative humidity, solar radiation, clothing level, and activity level. Scientists have been studying and using several human thermal comfort models for over 30 years based on both physiological and psychological responses [10]. Here, we review one widely used conventional thermal comfort model (PMV/PPD).

Conventional Thermal Comfort Models-PMV

The Predicted Mean Vote (PMV) is a widely used model for thermal comfort assessment. PMV was developed in 1960 by P.O. Fanger to represent the mean thermal sensation vote of a large group of people [10]. This model was created based on the difference between generated heat and released heat from the human body, and its correlation with the subjective perception of comfort. This approach is based on the heat-balance theory that relates cold discomfort to the mean skin temperature and warmth discomfort to skin sweat secretion. According to this theory, the neutral thermal sensation is the stage where the body produces heat and loses heat at the same rate. The scale of this subjective comfort is in a range from +3 for very hot to -3 for very cold and 0 for neutral.

These numbers deviate from the optimal presumed thermal comfort and were found by Fanger's study on healthy adult students. A comfortable environment is expected to have a PMV of 0 with a tolerance of ±0.5 units. The four contributing environmental factors of air temperature, mean radiant temperature, air velocity and relative humidity were gathered from the existing conditions, and two general tables were provided for the activity level and clothing insulation. Therefore, the PMV model is expected to *predict* the subjects' thermal sensation through knowledge of the aforementioned factors (activity level, clothing, air temperature, mean radiant temperature, air velocity and humidity). We can easily calculate the PMV through provided spreadsheets and online tools. There is no need to run the formula by the initial variables. These tools use formulas to convert the initial variables to measurable factors [1]. Although the PMV model has been used for buildings under the influence of mechanical heating, ventilation and air conditioning, the utilization was further extended throughout all building types and adopted by several standards such as ISO 7730, ASHRAE Standard 55, and CEN CR 1752.

As the PMV model was developed in a chamber setting under the air conditioning of the space, the result is expected to be different in actual settings and naturally ventilated buildings [8]. Previous research supports this model for higher quality performance results in buildings with natural ventilation [19], but has remained a debatable subject as some researchers believe that this quality of performance is simply a result of narrower thermal ranges in air-conditioned buildings [4]. The extended PMV model was introduced in 2002 to solve this problem in naturally ventilated buildings that are located in warm climates through consideration of the expectancy factor which relates the outdoor conditions and adaptive capability to the previous formula [6]. Since then, several research studies have been conducted to evaluate the reliability of the PMV factor and compare it with field studies in actual buildings or laboratory settings. Some studies have examined adaptive behavior as one reason for PMV-PPD prediction errors that provide the opportunity for adaptation to the existing conditions such as ventilation strategies and climatic conditions [5]. The adaptive behavior of the occupants is also responsible for the difference of PMV chamber results and actual buildings field studies [9]. Some of these studies have used the Observed Thermal Sensation (OTS) and Observed Mean Vote (OMV) to compare the accuracy of PMV-PPD both for individuals and groups of people. A PMV model was originally developed to predict the thermal sensation of groups of people however, the prediction accuracy for groups of people was not acceptable either in several studies [2]. The only acceptable prediction in the PMV model was in neutral conditions within the range of ±0.25, as bias was shown in both sides of the cool and hot sensation, with poorer performance in the cool side [4, 2]. In several studies, the PMV factor has shown poor performance both in individual level or group level prediction of the thermal sensation compared to observed thermal sensation. The main reason for this inaccuracy is several differences in the individuals that result in different thermal preferences which were not accounted for in the PMV calculation. The low accuracy of the PMV model brings concerns about using it in controlling our buildings as well.

Climate Appropriateness

It is important to note that not all materials can serve as effective enclosure systems. Furthermore, the façade system will vary depending on the region and specific climate conditions.

There is a drastic temperature difference between day and night in diurnal climates.

The seven climate zones are:

Zone 1: very hot

Zone 2: hot

Zone 3: warm

Zone 4: mixed

Zone 5: cool

Zone 6: cold

Zone7: very cold

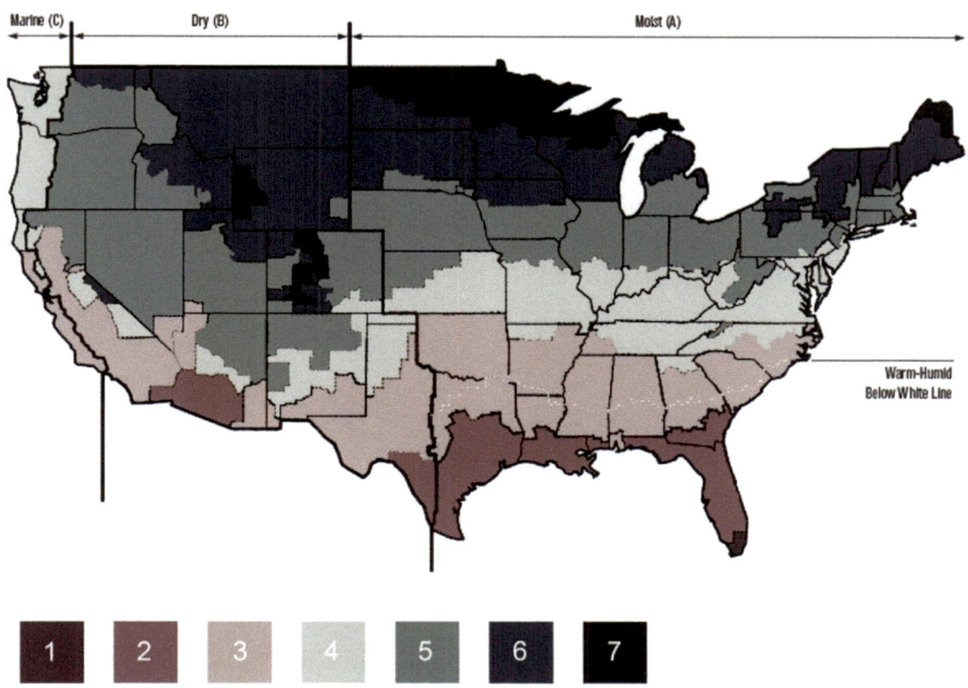

Figure 4.3 Climate Zones in the US

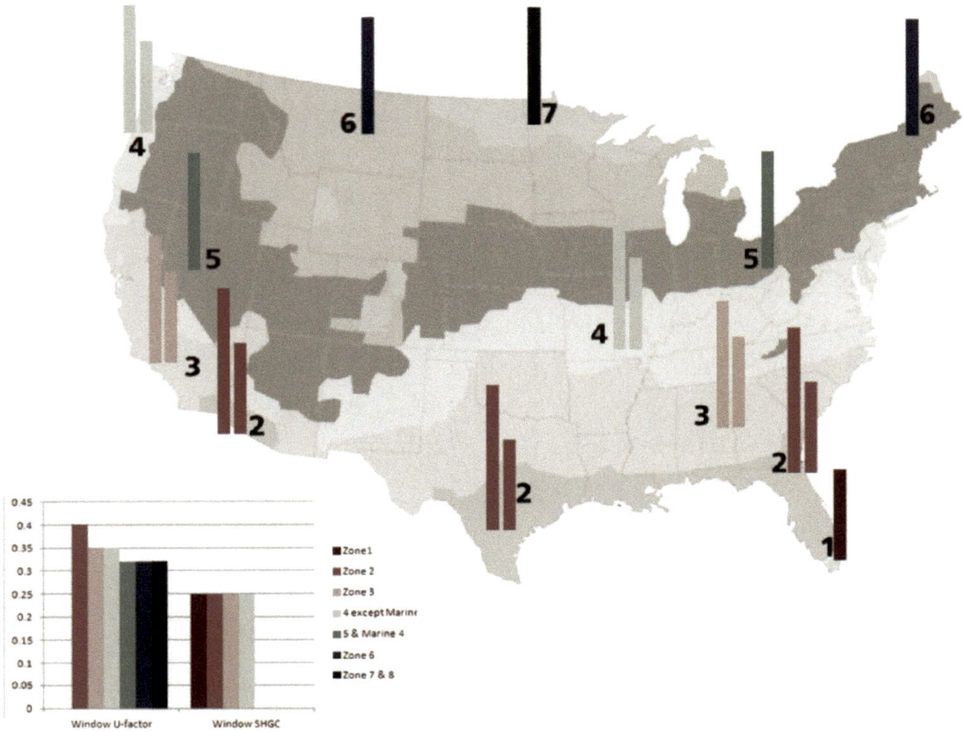

Figure 4.4 Minimum U-Factor and SHGC for Different Climates

In humid climates where the temperature tends to remain more constant over a 24-hour period. In such cases, additional methods of heating and cooling are needed to ensure an appropriate temperature range.

ASHRAE standards divide the US climate zones into seven categories (labeled by numbers 1 to 7) and three subzones based on humidity levels (marked by letters A: Humid, B: Dry, and C: Marine).

It should be noted that this classification doesn't recognize the microclimates that occur within the existing zones.

Context-Based Design Strategies

ASHRAE standards provide guidelines for heat transfer coefficients (U-Value) of the façade system as well as the Window-to-Wall-Ratio and Solar Heat Gain Coefficient (SHGC) for the glazed portion of the façade.

The minimum SHGC recommended by ASHRAE 2009 is 0.25 for all climates 1–7, and the U factor for the glass façade system should be less than 0.4 in zone 1 and 0.3 in zone 5, 6, and 7.

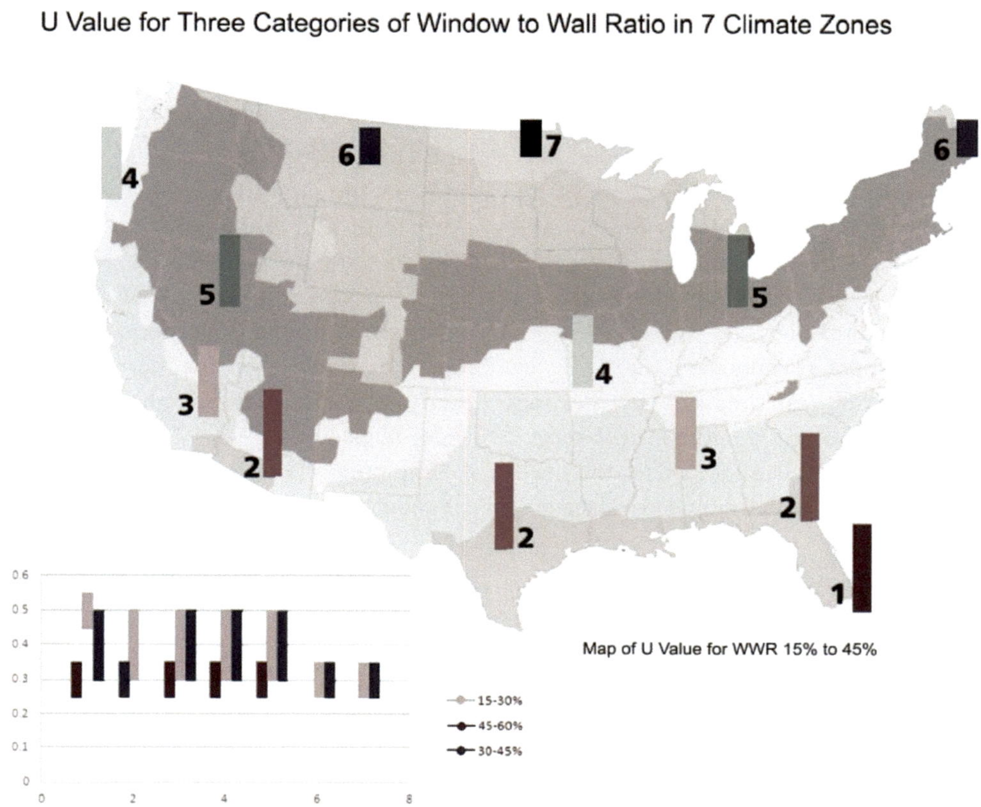

Figure 4.5 U-Values for Three Categories of Window Wall Ratio in Different Climate Zones

Double Skin Façade in Different Climates

New enclosures that can significantly enhance energy performance by allowing natural ventilation is a promising development. Aesthetics aside, a double skin façade (DSF) reduces cooling loads and allows for more or better natural ventilation, which is particularly notable in high-rise buildings. However, the performance of these façade systems should be considered for climates other than Western Europe. The goal is to clarify the performance of DSFs based on several research projects in this area so that designers can assess the value of these building concepts in meeting design goals for energy efficiency, ventilation, productivity and sustainability.

Temperate or Cold Climate

In temperate climates, a reduction in solar energy transmission is desired, so double skin façades are considered to be preferable. Their orientation has a significant impact on comfort and should be designed with utmost care. In the south-facing orientation, with an overcast sky, the cavity risks overheating. In office buildings, the heating needs are substantially less than the cooling, and research by [3] shows that shading is needed even on cold, sunny days.

In temperate climates, the DSF in buffer mode can reduce the heating energy significantly. If the heat is recovered at the top, it might offset the need in comparison with a single skin façade.

The results of calculations made by [10] between a double skin façade and a single skin façade show ranges from about 29% to 35% reduction of energy. In 2015, [7] looked at all the publications on the performance of the double skin façade and found out that, on average, the double skin façade creates a 33% reduction in heating load and 28% on the cooling load. Furthermore, their research shows that the narrower the cavity, the efficiency is higher than the wider ones.

The main concern for the winter months is to create a buffer zone to

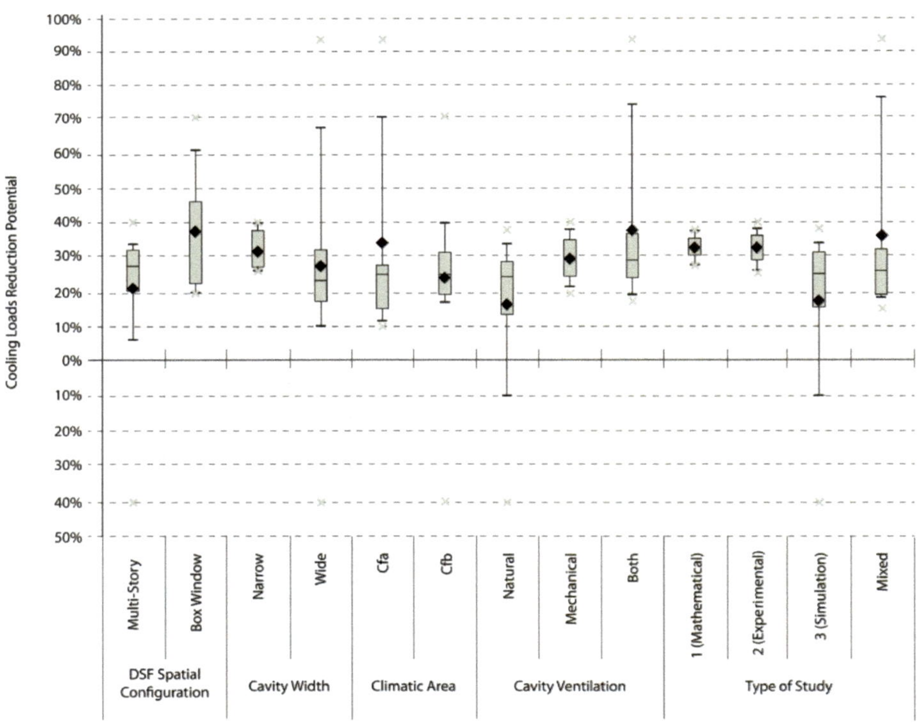

Figure 4.6 Cooling Loads Reduction Potentials

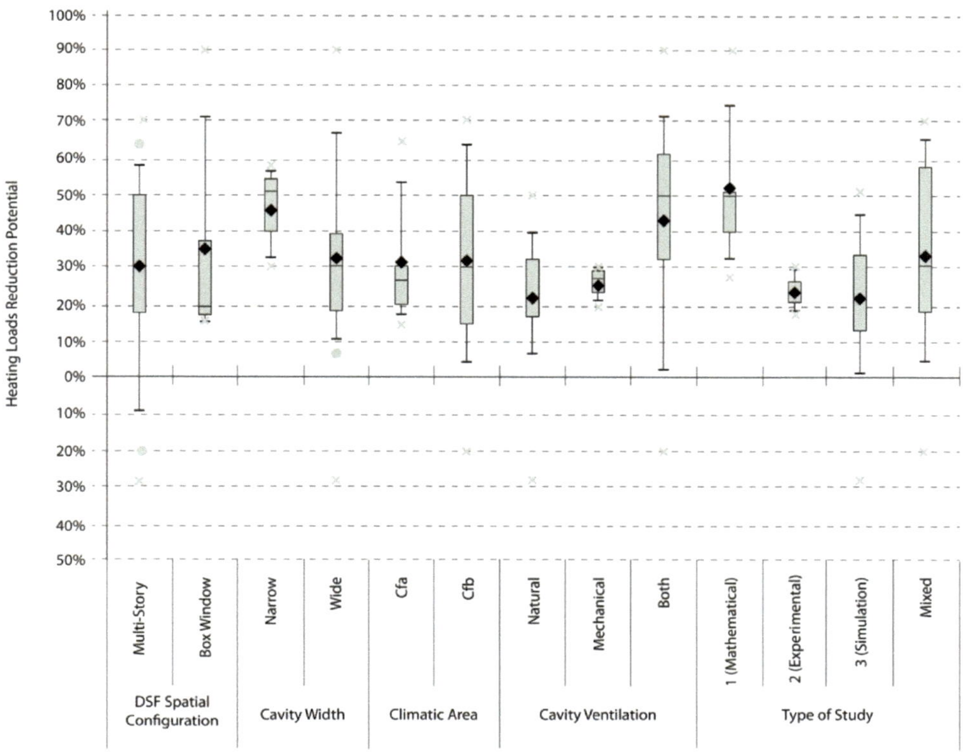

Figure 4.7 Heating Loads Reduction Potentials

minimize heat loss and reduce heating loads in cold climates. In addition, the DSF must balance the cooling loads for the summer months. Therefore, during the summer months, the cavity should be ventilated to protect from overheating.

Hot Arid

In hot and arid climates, the main concern is the cooling load. DSF in hot, arid climates is predicted to reduce the cooling loads compared to single skin reflecting façade systems. The cavity size should be narrow (more than 20 cm and less than 1 m) in order to prevent overheating. DSF can reduce the direct solar energy and result in a reduction of the cooling load, respectively. The lowest possible solar heat gain coefficient for the outer skin should be used in order to stop the heat from passing through the façade.

Regarding the type of ventilation, mechanical ventilation is more suitable for hot, arid climates than natural ventilation.

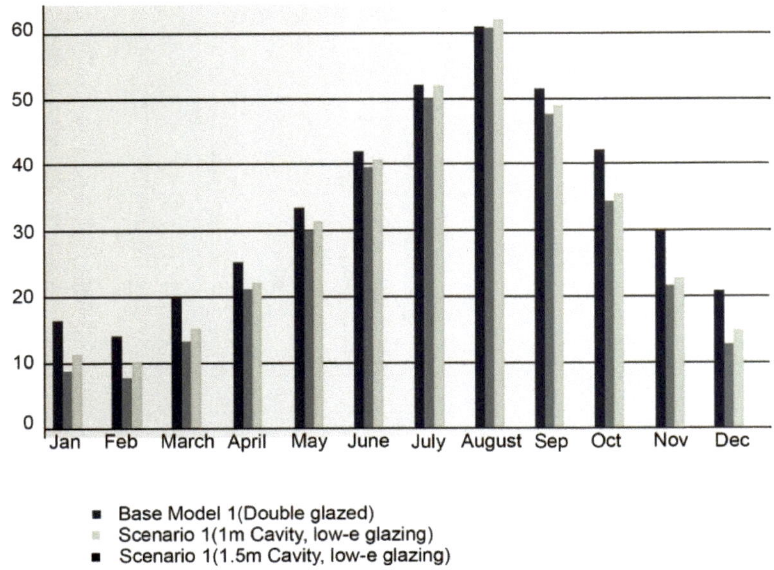

Figure 4.8 Total Energy Comparison of DSF vs. Single Skin in Hot and Arid

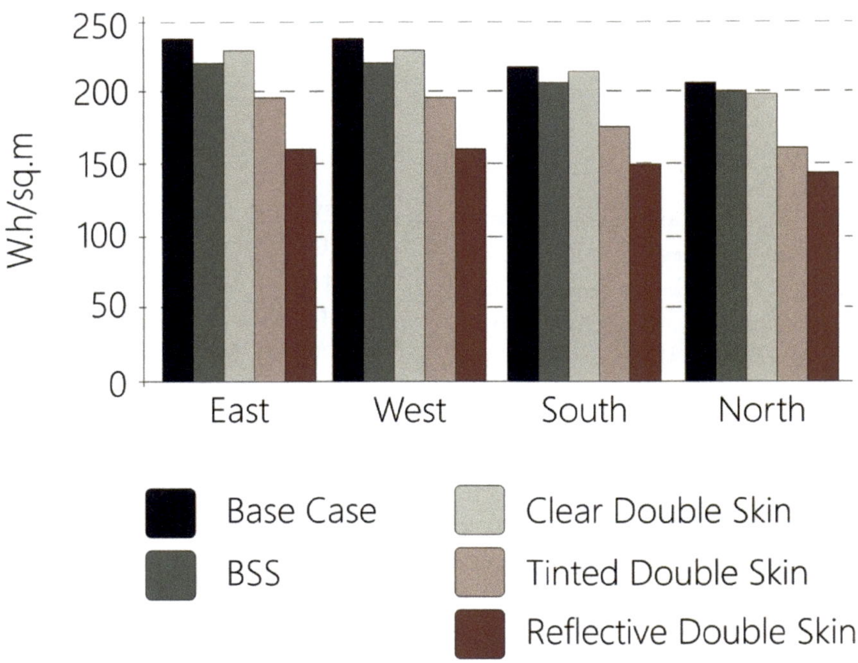

Figure 4.9 Peak Summer Total Cooling Loads Benchmark Single Skin Façade (BSS) Uses 40% WWR, Reflective Glazing and Selective Double Skin Façades

Hot Humid

The primary concerns for this climate type are to control heat transfer and solar radiation while increasing airflow. The façade might balance the requirements for ventilation and daylight while providing enough thermal protection through shading devices and glazing properties.

The shade plays a vital role in absorbing the heat and releasing it within the cavity of the DSF. Therefore, its position should be closer to the outer skin and not too close to trap the heat. Oesterle et al. suggested the best position would be the outer half of the cavity.

Regarding the height of the façade, research has shown 13 to 90 meters does not significantly impact the performance. However, the external curtain wall performs better than being an internal air curtain (or no airflow) in terms of the flow condition.

The results are shown in the graph in Figure 4.10

In addition, the same simulation suggests not to increase the width of the façade as that also has a negative impact on the performance.

In addition, the same simulation suggests not to increase the width of the façade as that also negatively impacts the performance.

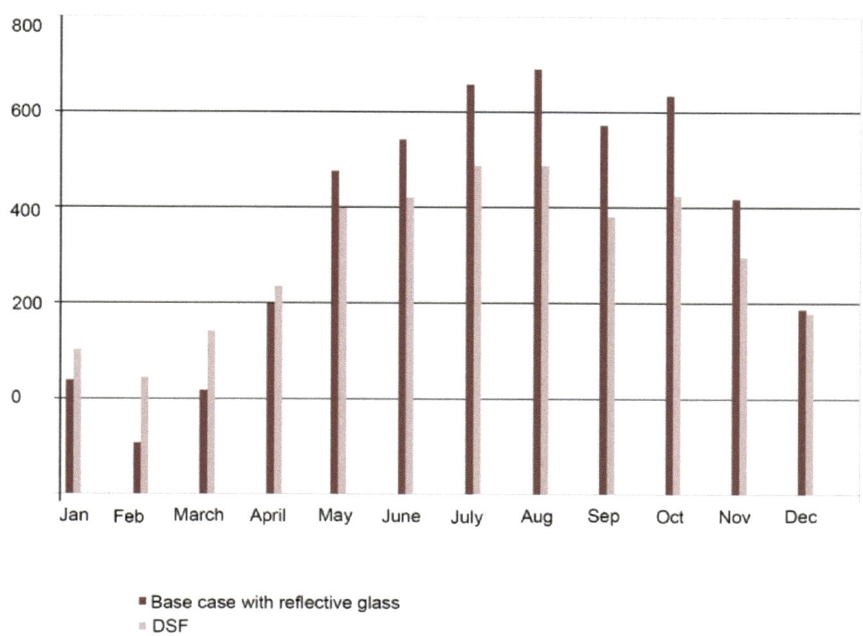

Figure 4.10 Energy Performance of Double Skin Façade and Base Case with Reflective Glass

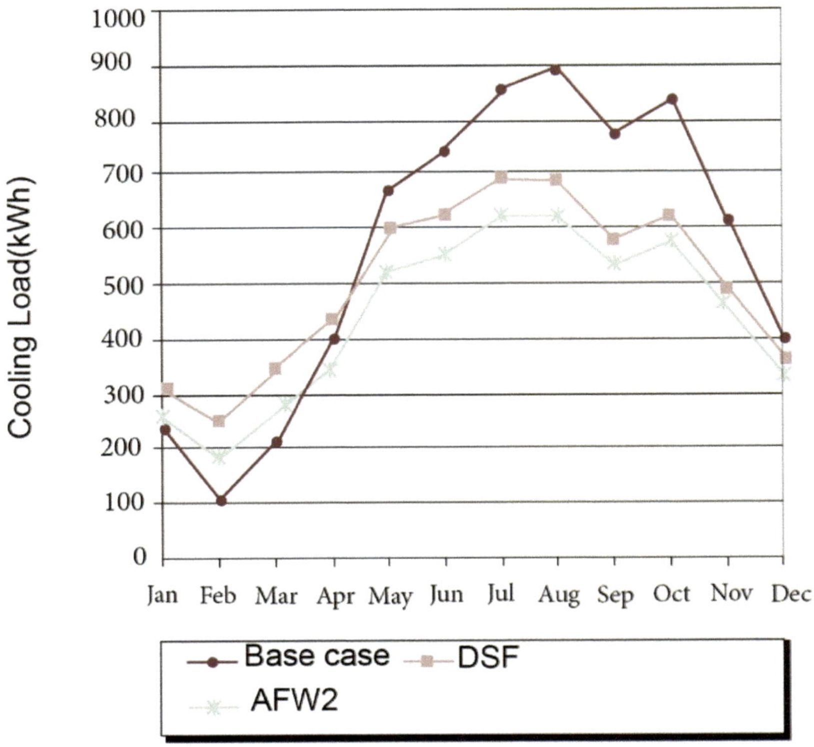

Figure 4.11 Cooling Loads of Double Skin Façade and Base Case with Reflective Glass

Conclusion

Research on the performance of the double skin façade in different climates revealed that varying scenarios should be considered. For instance, in hot and arid climates, where the primary concern is the cooling load, having a smaller size cavity with integrated shading devices proves necessary. The larger cavity heats up substantially and is challenging to keep cool. In temperate climates, the double skin façade is preferable, and narrower ones, in particular, demonstrate better performance than the wider ones.

In hot and humid climates, the shading position is quite essential; installation on the outer half shows better performance. In addition, securing a healthy airflow is required for this climate type.

References

1. Carlos, M., & Da Silva, G. (2013). *SPREADSHEETS FOR THE CALCULATION OF THERMAL COMFORT INDICES PMV AND PPD.* doi:10.13140/RG.2.1.2778.0887
2. Cheng, Y., Niu, J., & Gao, N. (2012). Thermal comfort models: A review and numerical investigation. *Building and Environment*, *47*(1), 13–22. doi:10.1016/j.buildenv.2011.05.011
3. Gelesz, A., & Reith, A. (2015). Climate-based performance evaluation of double skin facades by building energy modelling in Central Europe. *Energy Procedia*, *78*(C), 555–560. doi:10.1016/j.egypro.2015.11.735
4. Humphreys, M. A., & Fergus Nicol, J. (2002). The validity of ISO-PMV for predicting comfort votes in every-day thermal environments. *Energy and Buildings*, *34*(6), 667–684. doi:10.1016/S0378-7788(02)00018-X
5. Mishra, A. K., & Ramgopal, M. (2013). Field studies on human thermal comfort - An overview. *Building and Environment*, *64*, 94–106. doi:10.1016/j.buildenv.2013.02.015
6. Ole Fanger, P., & Toftum, J. (2002). Extension of the PMV model to non-air-conditioned buildings in warm climates. *Energy and Buildings*, *34*(6), 533–536. doi:10.1016/S0378-7788(02)00003-8
7. Pomponi, F., Piroozfar, P., & Pomponi, F. (2015). Double skin facade (DSF) technologies for UK office refurbishments: A systemic matchmaking practice. *Structural Survey*, *33*(4/5), 372–406. doi:10.1108/SS-04-2015-0025
8. Rupp, R. F., Vásquez, N. G., & Lamberts, R. (2015). A review of human thermal comfort in the built environment. *Energy and Buildings*, *105*, 178–205. doi:10.1016/j.enbuild.2015.07.047
9. Yang, Y., Li, B., Liu, H., Tan, M., & Yao, R. (2015). A study of adaptive thermal comfort in a well-controlled climate chamber. *Applied Thermal Engineering*, *76*, 283–291. doi:10.1016/j.applthermaleng.2014.11.004
10. Zerefos, S. (2007). On the performance of double skin facades in different environmental conditions. *International Journal of Sustainable Energy*, *26*(4), 221–229. doi:10.1080/14786450701803239

Bibliography

Cheung, T., Schiavon, S., Parkinson, T., Li, P., & Brager, G. (2019). Analysis of the accuracy on PMV – PPD model using the ASHRAE Global Thermal Comfort Database II. *Building and Environment*, *153*, 205–217. doi:10.1016/j.buildenv.2019.01.055

Haase, M., & Amato, A. (2008). Ventilated airflow windows for comfort in a hot and humid climate. *International Journal of Ventilation*, 7(2), 113–124.

Haase, M., Amato, A. (2009). A study of the effectiveness of different control strategies in double skin facades in warm and humid climates. *Journal of Building Performance Simulation*, *2*(3), pp. 179–187.

Haase, M., da Silva, F. Marques and Amato, A. (2009). Simulation of ventilated facades in hot and humid climates. *Energy & Buildings*, *41*(4): 361–373.

Hamza, N. (2008). Double versus single-skin facades in hot arid areas. *Energy & Buildings*, *40*(3), 240–248. doi:10.1016/j.enbuild.2007.02.025

McClain, J., Sifferlen, G., & Stoppelmoor, W. (2011). ASHRAE 90.1 reduces energy consumption: ANSI/ASHRAE/IES Standard 90.1-2010, Energy Standard for Buildings Except for Low-Rise Residential Buildings, provides guidance for reducing energy consumption by more than 20% over savings realized in the 2004 version. *Consulting Specifying Engineer* (January 31).

CHAPTER 5

Natural Ventilation and Double Skin Façade

Natural Ventilation Impact on Comfort

Natural ventilation is often deployed as a healthy option for space ventilation and may reduce the energy cost needed for air conditioning. For this reason, one must consider using natural ventilation, especially in light of the common emphasis on optimizing mechanical ventilation as opposed to natural. The aim of using natural ventilation is to take advantage of the natural forces as efficiently as possible, for as long a duration as possible, and to reduce the energy consumption used for cooling.

The development of more significant, more densely populated office buildings filled with computers, and designed around deeper floor plans that yield higher internal heat loads, has caused a shift toward airtight construction to better control the introduction of outside airflow and rely more on mechanical air conditioning. The use of natural ventilation in commercial office buildings is often restricted, even in temperate climates, by increased heat loads and concerns about occupant comfort. It was often believed that to achieve uniform internal temperature for occupant comfort, the only method was to control the amount of air supplied to an occupied space, particularly its temperature.

A well-designed building, constructed for natural ventilation, should be able to perform at a satisfactory level year-round. Summer climate conditions are of particular concern regarding occupant comfort. Therefore, during the summer months, the aim is to ventilate to meet desired comfort requirements and cool the space using purpose-provided openings.

As a primary focus, this research intends to evaluate the airflow patterns

DOI: 10.4324/9781315718750-5

and velocities that might be possible by natural ventilation in commercial office spaces during the shoulder season.

This chapter provides an overview of ventilation strategies, ranging from ventilation types to specific design considerations. First, we discuss types of ventilation: buoyancy-driven and wind-driven are presented. Finally, DSF design characteristics in naturally ventilated high-rise buildings are outlined, followed by a brief description of the reference building and its specific features.

Natural Ventilation Mechanism: Driving Forces

Two main forces cooperate to drive natural ventilation: Stack (or buoyancy-driven) ventilation and wind-driven ventilation. The specific properties of these two main forces are elaborated in the following section, where both individual and combined effects are described.

Stack Pressure: Thermal Buoyancy

Buoyancy-driven ventilation is prevalent in many naturally ventilated buildings, where pressure differences across the building envelope cause airflow. The air pressure differences here are due to temperature differences. The magnitude of the buoyancy of airflow is determined by the importance of the temperatures and the resulting differences in pressure. In stack-driven ventilation, height increases, and the pressure difference between an inlet and outlet is increased.

When the internal pressure is equal to the external force, a neutral pressure level (NPL) is created, which results in no airflow either in or out of an aperture at that specific height. Airflow and direction can be determined below or above the NPL and is always from the higher-pressure zone to the area of lower pressure. In other words, a temperature difference between the inlet and outlet can enhance the effects of buoyancy-driven ventilation. Air enters through openings in the lower part of the building and escapes through slots at a higher level when the inside air temperature is greater than the outside air temperature. Notably, the airflow direction is reversed when the inside air temperature is lower than the outside temperature. Stack pressure calculations are based on the temperature difference between the two air masses and the vertical spacing between the openings.

The NPL is calculated based on the total inlet and outlet areas and their respective resistances and relative height (if more than one-floor level exists). Roof openings and chimneys (or raised stacks) can shift the NPL, usually to higher levels. For buoyancy-driven flow, the NPL is presented graphically in Figure 5.1.

Here, the Bernoulli equation is used to derive the airflow from buoyancy-driven ventilation by calculating the pressure differential due to height — i.e., the hydrostatic head, for both the exterior and the interior environment. The overall pressure difference between the interior and exterior can be expressed in terms of the height difference, H, gravitational constant, g, density at a reference temperature, ρ_o, and the interior and exterior temperatures. The Bernoulli equation is given as:

$$\frac{v_O^2}{2}+\frac{P_O}{\rho}+gz_O=constant \tag{1}$$

About the buoyancy-driven case, there is a lack of external velocity, so the relationship is reduced to:

$$\frac{P_O}{\rho_O}+gz_O=constant \tag{2}$$

Using script E, the pressure difference is applied to the outside environment, and subscript I is used for the internal environment. The pressure difference that results from the height variation between an origin height, z_O, and at another height H, z_H, for the outside and inside become:

$$\Delta P_E=\rho_E g\left(z_H-z_O\right) \tag{3}$$

$$\Delta P_I=\rho_I g\left(z_H-z_O\right) \tag{4}$$

To determine the total difference in pressure, ΔP_T, the difference across the inlet and outlet apertures is determined. Illustrated in Figure 5.1, with the final resultant:

$$\left(P_{I,H}-P_{O,H}\right)+\left(P_{O,0}-P_{I,0}\right)=gH\left(\rho_O-\rho_I\right) \tag{5}$$

We presume air is a so-called "perfect gas." For this reason, we use the gas law:

$$\rho=\frac{P}{RT} \tag{6}$$

which is then substituted into the previous equation (5) for the terms ρ_O and ρ_I.

As the difference between the terms P_O and P_I is inconsequential when compared to atmospheric pressures, the term P/RT_B is set outside the parentheses, and again, we apply the ideal gas law.

$$\Delta P_T=\rho_O gH\left(\frac{T_I-T_O}{T_I}\right) \tag{7}$$

For ideal gases, we use the Boussinesq approximation where $\beta=1/T_I$. Concerning the Boussinesq approximation, differences in density are presumed to be inconsequential, except where needed to determine ΔP_T because the air density is not varied to any significant level concerning variation in temperature within the referenced building. As long as $\Delta T<30°C$, the Boussinesq approximation is largely accepted as valid.

The rate of ventilation is calculated by deploying the square root law [4], in conjunction with the Boussinesq approximation, while substituting in equation 7 we get:

$$\begin{aligned}Q&=c_d A\sqrt{2\frac{\Delta P}{\rho}}\\&=c_d A^*\sqrt{gH\frac{T_I-T_O}{T_I}}\\&=c_d A^*\sqrt{g\beta HT_I-T_O}\end{aligned} \tag{8}$$

Here, Q represents the flow rate through the building or space, C_d represents the coefficient of discharge, and A^* represents the contribution made by the inlet and outlet areas. A^* is defined as follows:

$$A^*=\sqrt{\frac{A_1^2+A_2^2}{A_1^2A_2^2}} \tag{9}$$

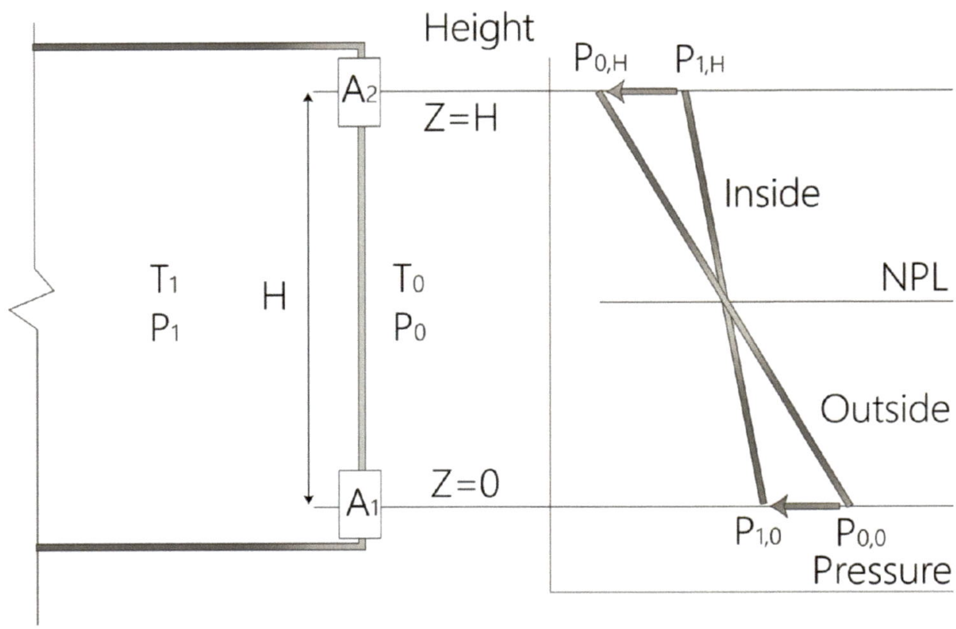

Figure 5.1 Neutral Pressure Level for Buoyancy-Driven Ventilation

The airflow is not assisted by forced air from wind or mechanical systems when natural ventilation is used in a buoyancy-driven case. Quite often, this presents a critical design issue for applying buoyancy-driven ventilation as a passive strategy during the warmer summer months. In a buoyancy-driven case, the following parameters are somewhat interdependent, making the analysis of this ventilation scheme more complicated. The parameters included:

- The size of the outlets and inlets
- The particular height of the space
- The strength of the heat sources that cause airflow
- Sources of interior heat cause the temperature that results from the difference between interior and exterior spaces.

Furthermore, complicated building geometries (for example, the several floors that are directly or indirectly involved) make it more challenging to analyze the forces that cause natural ventilation flow. The effect of natural ventilation in building design is reduced partly by this complex geometry in conjunction with the overall lack of understanding of the physics involved in the wind- and buoyancy-driven ventilation.

Wind-Driven Ventilation

Several environmental conditions influence natural ventilation, the most unpredictable of these being wind velocity, precisely, its speed and direction. These two factors are both difficult to control and analyze, particularly in a full-scale building. In the real-world environment, instantaneous wind speed varies with time, and pressure differences vary depending on building geometry and location on the surface of the building. A uniform and constant wind speed are used in most wind-driven natural ventilation experiments. These design wind speeds are often defined as the mean wind speeds for a given location over a specific period of time, typically over a few years or decades [1].

Several formulas were developed to illustrate differences in pressure caused by wind-driven flow. The formulas below describe a constant wind speed where wind pressure will not fluctuate over time. That said, with regard to single-sided ventilation, fluctuation in wind speed may prove critical. The diagram in Figure 5.2 below represents wind-driven ventilation, the direction of airflow, and the resulting pressure with relation to height.

In a situation where the apertures on opposite sides are identical, the differences in pressure across the apertures are equal to one-half the difference in pressure across the building when it can be assumed that the pressure differential through the interior of the building is negligible. The Bernoulli equation will then reduce to the ideal equation when applied between a point at a distance from the building face that contains the window and the building façade:

$$P_w = P_O + \rho_O \, U_O^2 / 2 \qquad (10)$$

where P_w is the pressure caused by wind on the façade and P_O represents the pressure at a distance from the building, and U_O is the reference of velocity also at a distance from the building. If any is detected, the differential in height is neglected for flow along a particular streamline. Thus both gz terms are zero, and the velocity at the building façade is also zero, as it is now

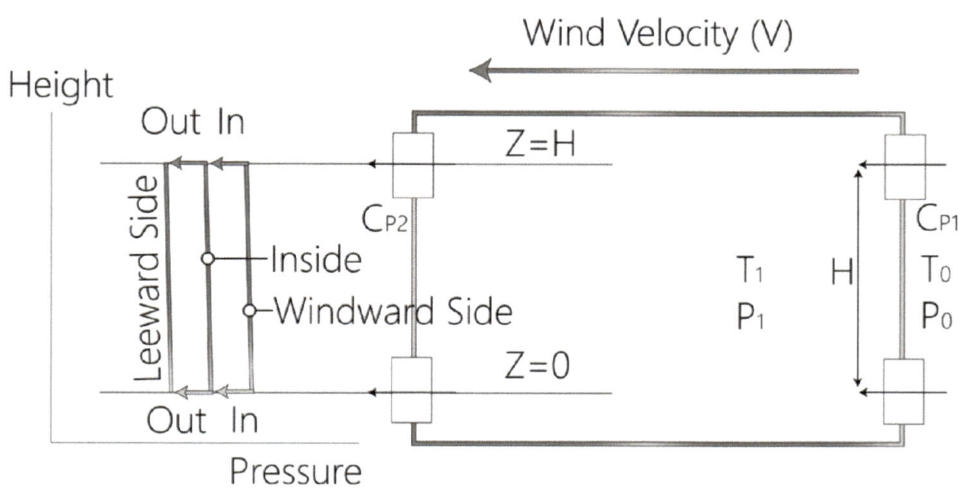

Figure 5.2 Wind-Driven Ventilation: Airflow Direction and Pressure Versus Height

the point of stagnation. The formulas that result are described as follows:

$$P_w = P_O + C_P \rho_O \frac{U_O^2}{2} \quad (11)$$

$$Q = C_d A U_O \sqrt{\frac{\Delta C_P}{2}} \quad (12)$$

Where Q represents the flow entering and exiting through the apertures, the value of c_p is dependent on the of the building as well as the façade's location; additionally, values are often discovered by deploying wind tunnel experiments [5]. It is assumed that pressure on the outside of the building in Figure 5.2 does not vary with height to any significant degree.

Thermal Buoyancy and Wind in Combination

A combination of wind and buoyancy flow is more commonly observed in full-scale buildings; the two forms of natural ventilation may work in tandem or in opposition. Figure 5.3 describes the direction of airflow and its relation to pressure compared with height in a combined wind-buoyancy ventilation setting. The total of the pressures, in this case, are added together to determine an opening.

$$\Delta P_T = \Delta P_W + \Delta P_B \quad (13)$$

The total rate of flow through an aperture is calculated using the law of square roots as described in formula eight as:

$$Q_T = C_D A \sqrt{2 \frac{\Delta P_T}{\rho}} \quad (14)$$

By substituting the differences in pressure for each case in formula 12, then the total pressure difference into formula 2.13, the total rate of flow, Q_T, becomes:

$$Q_T = C_D A \sqrt{\Delta C_P \frac{U_O^2}{2} + g\beta H (T_B - T_O)} \quad (15)$$

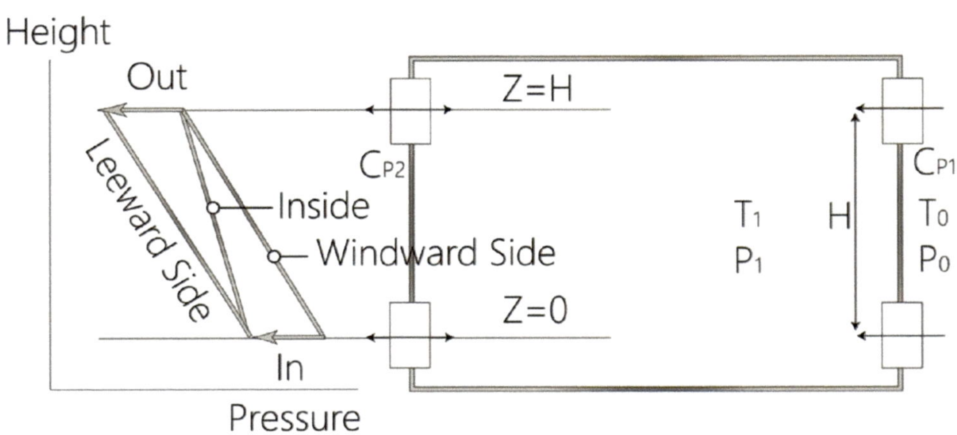

Figure 5.3 Combined Wind-Buoyancy Ventilation: Airflow Direction and Pressure Versus Height

$$Q_T = \left[Q_W^2 + Q_B^2 \right]^{1/2} \tag{16}$$

where Q_w is defined at the rate of flow component caused by wind while Q_B is the component associated with stack or buoyancy flow[1].

Concerns about this formula's accuracy do exist [4] partly as a result of the effects of wind and buoyancy. When the effects of wind and buoyancy are roughly equivalent, the error from using formula 15 was typically around 50% [4]. The level of the error caused by using formula 15 is dependent upon the distribution of both inlet and outlet apertures and characteristics of flow at those openings. Currently, on the other hand, there is a lack of availability of simplified methodologies to calculate the rate of ventilation where wind and buoyancy are active at the same time.

Thermal conditions play a part in driving air change; for this reason, a ventilation component in the energy balance of space must necessarily be included. With respect to a simple room, the total measure of airflow, ρcPQTΔT, added to the conduction of heat through the building's envelope must be equal to interior heat loads under steady-state conditions. The differences in temperature between indoor and outdoor environments, the surface areas, and material properties are deployed to calculate the rate of conduction through walls and windows: Qwalls+windows=UAΔT. The internal loads, or Qloads, usually account for the heat caused by lights, equipment, and occupants. The formula for energy balance in steady-state conditions is:

$$Q_{loads} = UA\left(T_I - T_O\right) + \rho\, c_P \left(T_I - T_O\right) \tag{17}$$

The schematic diagram (Figure 5.4, left) describes buoyancy-induced pressure distribution on the envelope of an oval-shaped high-rise building. The dotted lines pointing inward represent "underpressure." The solid ones pointing outward indicate "overpressure." As a result of the differences in interior and exterior temperatures, a differential in pressure over the envelope forms. The schematic diagram (Figure 5.4, middle) describes the wind-induced pressure distribution on the same building.

The dotted lines pointing inward describe the positive pressure that develops on the windward side of the building's envelope, while the solid lines that point outward describe the "underpressure" that emerges on the envelope's leeward side. Positive and negative pressures toward the top of the building both increase due to the wind profile. The schematic diagram (Figure 5.4, right) indicates the overall

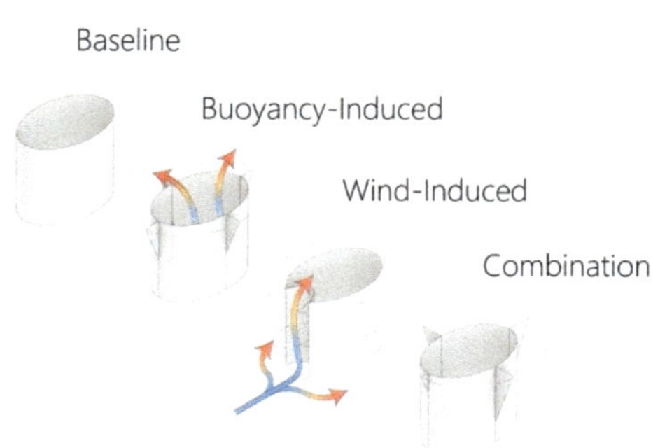

Figure 5.4 Left: Buoyancy-Induced Airflow. Middle: Wind-Induced Airflow. Right: Combination of Both Driving Forces

distribution of pressure differentials and the effect of wind buoyancy on the building's envelope. The figure demonstrates that pressure gradients calculated from wind and buoyancy forces can be summed. They serve to either neutralize or strengthen one another [3].

Natural Ventilation Principles

A building's shape, along with the location of the ventilation apertures, determines the manner of ventilation of the natural ventilation. There are generally understood to be three different principles of natural ventilation:

- Single-sided ventilation
- Cross ventilation
- Stack ventilation

These principles of ventilation describe the way in which the interior and exterior airflows are related, in other words, the manner by which the driving forces get deployed toward the ventilation of a building. The principles also show the ways air is brought into a building and how it is spent. Depending on how airtight the building is, infiltration through the building's envelope may also play a role. However, this form of ventilation is often unintentional and undesired.

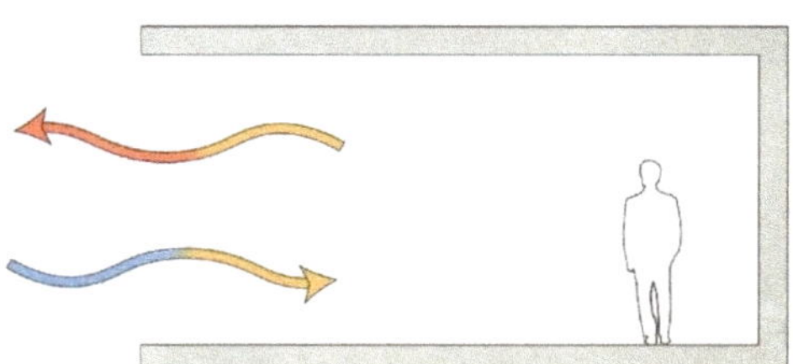

Figure 5.5 Sketch of the Single-Sided Ventilation System

Single-Sided Ventilation

Single-sided ventilation relies on opening(s) on only one side of the ventilated enclosure. Fresh air enters the room through the same side that used air is exhausted. Rooms within a cellular building with operable windows on one side and closed internal doors on the other side are an example of this type of ventilation.

As a rule of thumb, single-sided ventilation is effective to a depth of about 2–2.5 times floor-to-ceiling height [2]. With a single ventilation opening in a room, the main driving force in summer is wind turbulence. In cases where ventilation openings are provided at different heights within the façade, the buoyancy effect can enhance the ventilation rate. The contribution from thermal buoyancy depends on the temperature difference between the inside and the outside, the vertical distance between the openings, and where the openings are located. The greater the vertical distance between the openings and the greater the temperature difference between the inside and the outside, the more substantial the buoyancy effect. Compared with other strategies, lower ventilation rates are generated, and air does not penetrate far into the space.

Generally speaking, single-sided ventilation is adequate to a depth of roughly 2 to 2.5 times the floor-to-ceiling height [2]. Where there is only a single aperture for ventilation in space, the primary driving force during the summer is wind turbulence. The buoyancy effect may improve the ventilation rate in situations where ventilation apertures are built at varying heights relative to the façade. The contribution of thermal buoyancy is dependent on several factors: the difference in temperature between indoor and outdoor, the vertical distance between the apertures, and the locations of the apertures. The effect of buoyancy becomes more significant as the vertical distance between apertures and variation indoor/outdoor air temps.

Cross Ventilation

In cross ventilation, the opening are located across from each other of a space of building and can provide natural ventilation for the occupants. If designed propergy, it can remove the heat from a space and maintain indoor air temperatures around 1.5 C° (2.7 F°) above the outdoor temperature.

Stack Ventilation

So-called "stack ventilation" occurs as a result of driving forces that promote outflow from the structure. Typically, fresh air enters by way of low-level apertures, while contaminated, used air is exhausted through apertures at higher levels (note: a reversal of this flow can happen under certain conditions). The effectiveness of stack ventilation can be enhanced by designing the outlet to be situated in a wind-induced underpressure region.

A typical rule of thumb indicated that stack ventilation is most effective at a width of 5 times the floor-to-ceiling height when measured from the inlet to the place of exhaustion.

A typical example of this would be a building with an elevated central area where warm, contaminated air from the surrounding spaces rises up in order to be released through wind towers on the rooftop. Due to the laws of physics, the stack effect requires that a particular height between the inlet and the outlet be established. There are various methods available to achieve this goal, notably increasing the floor-to-ceiling height while tilting the roof's profile or including a chimney or atrium in the space. With regard to specific individual spaces, stack ventilation may resemble cross ventilation in that air enters on one side of the space while exhausting on the opposite side. The air may flow the entire width of the structure or from the edges

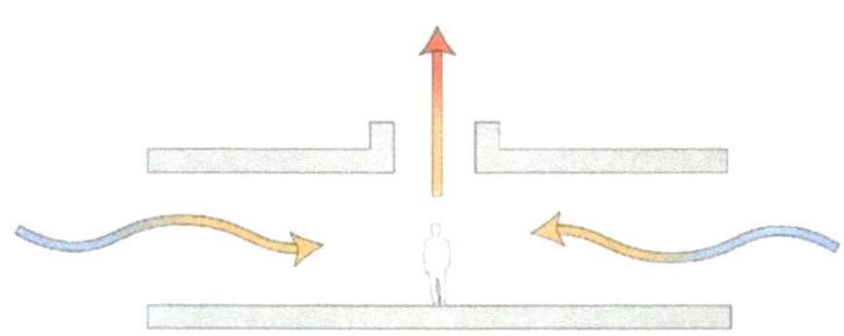

Figure 5.6 Sketch of Stack Ventilation

of the building to its center to be released by way of a central chimney or atrium.

Elements of Natural Ventilation

Certain characteristic elements enhance natural ventilation principles. These elements serve to distinguish the mechanisms of natural ventilation from other mechanisms of ventilation. Furthermore, they illustrate effects on architecture, economy, and performance.

Interestingly, natural ventilation can be achieved without the use of these dedicated elements. For example, the structure itself can double as an element for ventilation. This element could be called a "building-integrated element." As such, the building would be designed with the goal of harnessing the natural forces and directing ventilating air through the spaces without dependency on specified ventilation elements. In other words, a building-integrated element is not actually functioning as one; rather, it is the absence of an element. Since the ventilation system and the building's occupants share the same spaces, and since windows and doors are deployed as critical parts of the airflow paths, the most characteristic feature of this type of system is the absence of an actual system.

Neither mechanical nor traditional natural ventilation elements exist here. The primary advantage of a building-integrated system would be that it requires no additional consumption of space. The majority of naturally ventilated buildings do, however, require the use of dedicated elements for ventilation to take advantage of natural driving forces and encourage airflow through the structure. Described in the following section are the elements used for this research. Included are descriptions of the advantages and disadvantages of the individual element system.

Double Façades

The addition of a second glazed envelope is used in a double façade, thus creating an opportunity for maximum daylight and improvement of energy performance. A double façade shares many properties of an atrium. However, the cavity of a double façade cannot be occupied.

A natural ventilation double façade can be deployed as a path for an inlet or outlet in any of the three natural ventilation principles. Many significant advantages are offered, such as:

- The cavity is guarded against outside noise and wind. Therefore, windows can be open regardless of wind and noise outdoors, even on the higher floors of the high-rise structure.
- When located within the cavity, devices for solar shading are protected from the wind.
- When the cavity is deployed as an air supply path, solar preheating of the air supply is provided on sunnier days.
- Transmission losses through the wall are decreased relative to traditional external walls due to the protective nature of the cavity.
- As a result of the climate protection in the cavity, indoor window surfaces are

warmer, thus reducing frigid downdrafts and asymmetric radiation.

Disadvantages:

- High cavity temperatures can be a problem on hot days if the external glazing is not operable. This is mainly a problem in the upper floors of a building when the double façade is used as an exhaust stack.
- Noise can be transferred between adjacent rooms by reflection in the glazed cavity surfaces.
- Cleaning of the cavity is important, mainly when used as a supply air path. This implies higher operation costs than a normal façade.
- A double façade represents significantly higher construction costs than a normal façade; however, because they are usually not built solely for ventilation purposes, the costs can be distributed for other functions, e.g., daylight and visual amenity.

Figure 5.7 The Queens Building (Peake Short & Partners, 1993) at Leicester University with Chimneys Extending Well Above Roof Level to Increase the Stacking Potential and Achieve Stable Wind-Induced Suction

Credit: John Lord

Chimneys

Another common element used in natural ventilation systems is cylindrical chimneys. The majority of chimneys provide an increased buoyancy effect while extracting the ventilation air. Because of a chimney's strategic positioning high above the roofline, the opening is conducive to strong, stable winds and has the potential to act as a driving catalyst for ventilation. In order to maximize a building's underpressure, careful consideration should be made when designing the openings of the chimney. An open top provides the most straightforward design and ensures negative pressure and suction no matter the direction of the wind, as a result of the Bernoulli effect. In order to prevent rainwater ingress, a cover top can be affixed. As an alternative, a design that combines roof cowls may ensure a greater level of protection from the weather while increasing the chimney's effect.

A key advantage of using a chimney is that they provide a simple and efficient way to use both thermal buoyancy and wind, completely independent of the direction of the wind.

Natural Ventilation and High-Rise Buildings

A building that is taller than ten stories is considered a high rise. Several low-energy innovations such as natural ventilation can be integrated into a high-rise building. For this reason, the high-rise is of particular interest and constitutes the focal point of this research.

The most important element of the low-energy concept is the façade itself. A higher degree of transparency ensures maximum exposure to daylight and exterior views. Here, the control of heat and lights is obtained by the use of shutters and blinds. Further, the double façade's "buffer zone" also contributes to improved insulation values. The most important factor to consider when exploring high-rise ventilation options is that the building's velocity profile increases with height. The most conventional method used to approach this problem has been to seal the façade and place within it a mechanical-ventilation plant. As a result of the building's height, windows cannot be opened at that level, and thus, there is no shading because they would otherwise flap around in the wind. Furthermore, a building's location in a bustling urban environment presents other issues associated with an open window, such as traffic noise and air pollution. Therefore, our concept was to close the façade with the second skin and ventilate the offices into a double façade.

Naturally Ventilated Double Skin Façades

With regard to buildings with double skin façades, various interesting investigations have been published in the literature about passive ventilation. It was discovered that significant savings of energy are indeed possible if natural ventilation is exploited by the use of a double skin façade (DSF). For example, the Loyola University Information Commons and Digital Library in Chicago integrated natural ventilation with a DSF to cool the buildings. The results indicated that an integrated façade could reduce energy consumption by 30% (when operated for 52 days in natural mode) (Frisch, 2005).

Despite the fact that most of the research has been done under temperate climate conditions, studies have indicated a close link between the design of natural ventilation and the function of DSF. A simulation algorithm was developed to investigate the behavior of temperature characteristics of the flow of natural convection DSFs through solar radiation [6].

It was discovered that air temperature increased close to heat sources, which were near window panes and shading devices. Strategies for natural ventilation, greenhouse effects, as well as the best positioning for sun-shading devices were investigaged on south-facing DSFs in temperate climates of the Northern hemisphere [7–14]. They discovered that an adequate rate of day/night ventilation might be achieved by a window opening even where the wind characteristics were unfavorable. They further argued that with regard to a south-facing DSF, during the cooling season, when air flows downward as a result of higher wind pressure at upper openings, a danger exists when using air in the cavity to ventilate the building. With this situation, warmed air in the cavity is pulled into the building, thereby adding to the cooling load.

The air temperature of the cavity is always higher than the outdoor air during daytime hours when there is incident isolation. In the case of a south-facing DSF, this suggests that a sophisticated louver control system is needed to assure that cooling loads won't increase due to natural ventilation. Thus, special care must be taken to work against greater wind pressure at the top cavity opening than those at the bottom. Airflow within the cavity is typically dependent on wind pressure as opposed to buoyancy on a windward façade; thus, without the use of a careful design for prevention, the air will indeed flow downward. If ventilation is the primary goal, this phenomenon could lead to the use of a northern-oriented DSF. Additionally, the authors include general guidelines for the improvement of natural ventilation in office buildings with a DSF during daytime and effectively proved that efficient, natural cross-ventilation is possible in climate conditions that are temperate.

Conclusion

Buoyancy and wind-velocity are the two primary forces by which fresh outdoor air can be channeled into an interior space. At the same time, unwanted radiant heat (solar radiation and internal heat gains) is expelled externally. Though these principles are primarily understood, more research is needed to explore the full extent of DSF's capacity to reduce energy usage. Specifically, the potential exists for a shaft-type façade system, which will reduce energy consumption in high-rise buildings by deploying natural ventilation strategies.

References

1 Awbi, H.B. (2003). *Ventilation of buildings*, Routledge.
2 CIBSE – Chartered Institution of Building Services Engineers (1997). Natural ventilation in non-domestic buildings. Chartered Institution of Building Services Engineers.
3 Daniels, Klaus. (1998) Low-tech light-tech high-tech: building in the information age. Birkäuser.
4 Etheridge, D., & Sandberg, M. (1996) *Building ventilation: Theory and measurement.* John Wiley & Sons.
5 Orme, M. (1999). Applicable Models for Air Infiltration and Ventilation Calculations, AIVC Technical Reports.
6 Von Grabe, J. (2002). A prediction tool for the temperature field of double facades. *Energy & Buildings, 34*(9), 891.
7 Gratia, E., & De Herde, A. (2007a). Are energy consumptions decreased with the addition of a double-skin? *Energy and Buildings, 39*(5), 605–619.
8 Gratia, E., & De Herde, A. (2007b). Greenhouse effect in double-skin facade. *Energy and Buildings, 39*(2), 199–211.
9 Gratia, E., & De Herde, A. (2007c). Guidelines for improving natural daytime ventilation in an office building with a double-skin facade. *Solar Energy, 81*(4), 435–448.
10 Gratia, E., & De Herde, A. (2007a). The most efficient position of shading devices in a double-skin facade. *Energy and Buildings, 39*(3), 364–373.
11 Gratia, E. (2004a). Is day natural ventilation still possible in office buildings with a double-skin facade? *Building and Environment, 39*(4), 399.
12 Gratia, E., & De Herde, A. (2004b). Natural cooling strategies efficiency in an office building with a double-skin facade. *Energy & Buildings, 36*(11), 1139–1152.
13 Gratia, E., & De Herde, A. (2004c). Optimal operation of a south double-skin facade. *Energy & Buildings, 36*(1), 41.
14 Gratia, E., and De Herde, A. (2004d). Natural ventilation in a double-skin facade. *Energy and Buildings, 36*(2), 137–146.

CHAPTER 6

Performance Analysis and Case Studies

An important element of the book are the interviews with different firms involved in design or construction of double skin façade projects. Eight double skin buildings in the US have been studied. All these case studies are commercial buildings in different climatic contexts and presented in Chapter 6.

The goal of the interviews was to distinguish between the different types of double skin façades in different contexts (from a structural and construction point of view) and find out what the purpose was for the construction of double skin façades, their choice of different systems, their benefits and addressing the code in regards to fire control.

Questions were asked concerning the following aspects:

Why are double skin façades constructed and what are their benefits?

- Design
- Dimensions
- Types of construction
- Sun shading
- HVAC issues
- Operation
- Ventilation aspect

In double skin façade projects, the architect teams carried out the design and chose the construction types of the system while a separate consultant studied the performance of the façade system, and in many cases, they hired separate contractors who erected it. Five major designers of double skin façade including contractors

DOI: 10.4324/9781315718750-6

and building operators were interviewed and their experiences are listed below: Architects distinguish between the façades constructed in the US and the ones in Europe. In the US, all the façades are multi-story façades and all of them shared the following operational benefits:

- The cavity acts as a thermal buffer zone during wintertime and it was cooled during shoulder season and summer-time due to stack effect and the two openings at the bottom and the top of the façade.
- Sun shading placed in the cavity protected from the exterior environment

In other countries, such as Germany and England, different types of façades, such as corridor type and box types, are constructed, of which we don't have any examples in the US. The benefits associated with those designs are that the rooms can be ventilated naturally and their outer skin opening is not limited to only two (top and bottom) as it is in the US.

Figure 6.1 Cambridge Public Library

CAMBRIDGE PUBLIC LIBRARY

LOCATION
Massachusetts, Boston

BUILDING PROGRAM
Public Library

ARCHITECT
William Rawn Associates

SIZE
27,200 SF

COMPLETION DATE
1990

BACKGROUND AND CONTEXT

The Cambridge Public Library is 104,000 SF, of which 76,800 SF, are for the new building and 27,200 SF for the historic building (Van Brunt & Howe, 1889). The new building is a project which incorporates a double face system; it is a pioneer of double skin curtain wall technology in the US. This 76,800 SF addition of a new construction to the Cambridge Public Library in Boston employs a multistory double skin façade. Nestled in the historic heart of Boston, the new addition to the Cambridge library is an architectural bridge that links the traditional past to a visionary future. The building will not only conserve significant amounts of energy, but it will also create a highly conducive environment for study and research while ushering in a new era of campus architecture focused on resource conservation.

Figure 6.2 Building on the Site

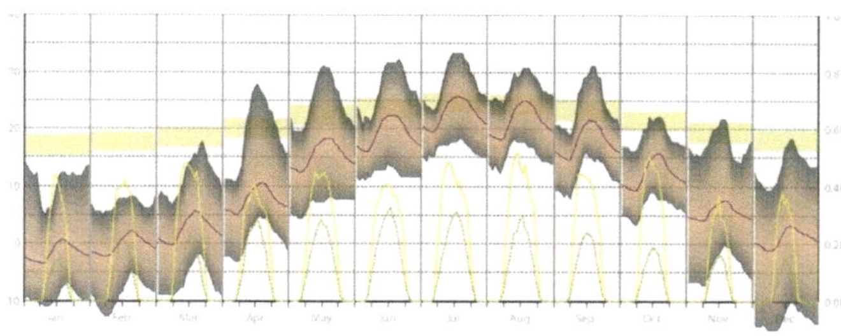

Figure 6.3 Hourly Temperature

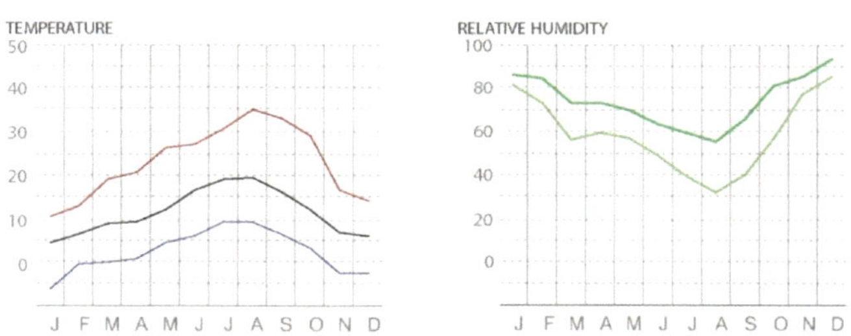

Figure 6.4 Climatic Data – Monthly Temperature, Relative Humidity

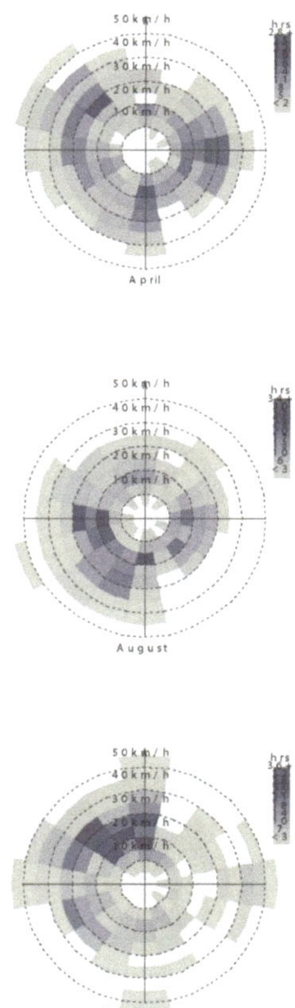

Figure 6.5 Prevailing Winds

The climate is classified as a humid continental climate (Koppen Dfb with areas of Dfa). Summers are typically warm, rainy and humid, while winters are cold, windy and snowy. Spring and fall are usually mild, but conditions are widely varied, depending on wind direction and jet stream positioning. The hottest month is July, with a mean temperature of 73.9 °F (23.3 °C). The coldest month is January, with a mean of 29.3 °F (–1.5 °C).

The city averages 42.5 inches (1,080 mm) of precipitation a year, with 41.8 inches (106 cm) of snowfall a year. Prevailing wind directions throughout the year are South-West and North-West.

DESIGN INTENT AND BUILDING FEATURES

Some specifications of the project are:

- 3'-0" Deep Airspace: provides added insulating depth and enables maintenance access
- Multi-story Double Skin: the greater height improves natural convection and makes the heat capture and exhaust more efficient.
- Movable 1'-0" Deep Sunshades: locating the blinds in the protected cavity allows them to collect the sun's heat energy before it enters the conditioned building. The lightweight aluminum louvers can be motorized to provide glare protection at all sun angles (important on a southwest exposure).

A Unique High-Performance Façade:

- The double skin façade (two surfaces of glass, and a cavity, creating insulated airspace) is a multi-story (full height), full depth (3'), thermal flue. The façade allows for complete transparency while ensuring protection from excessive heat gain, heat loss, and glare:

Energy Savings and Comfort

- The façade saves energy (50% reduction compared with a conventional curtain wall) and maximizes comfort in the reading spaces. The 3' airspace can be open in summer to keep heat from entering the building and closed in the winter to create an insulating "thermal blanket."

Natural Light

- The façade brings a significant amount of balanced natural light into the library; carefully controlled by the fixed and movable sunshades.

Natural Ventilation

- Operable windows in the façade allow for fresh air throughout the year (even in winter) without insect screens blocking and without concern for stolen books. In the winter, spring and fall, the windows allow heat from the cavity to be brought into the building.

The southwest-facing double skin façade – by Gartner Steel and Glass GmbH, based in Würzburg, Germany – consists of inner and outer glass walls separated by a 3-ft deep by 40'-6" tall cavity. The two skins are supported by a series of vertical steel frames tied back to the building's primary structure. The framing supports the point clamped monolithic exterior skin, as well as large translucent glass shading canopies on the exterior that act as light shelves, aluminum blinds within the cavity, and the aluminum curtain wall interior skin utilizing insulated glass units.

The façade allows for complete transparency while ensuring protection from excessive heat gain, heat loss, and glare. The façade saves energy (50% reduction compared with a conventional curtain wall) and maximizes comfort in the reading spaces. The 3' airspace can be open in summer to keep heat from entering the building and closed in the winter to create an insulating "thermal blanket." Operable windows in the façade allow for fresh air throughout the year. In the winter, spring and fall, the windows allow heat from the cavity to be brought into the building. In the summer, the windows allow natural ventilation into the building.

The building has three floors and a basement, and the zones are divided according to their use. We have eight different zones: hall/lecture/assembly area, circulation area, display and public area, generic office area, reception, toilet, workshop-small scale, light plant room and cupboard. The plan was divided into multiple zones according to geometry, proximity, and use. Each zone was then assigned an activity template (indicated by color in images) which associates that zone with different energy profiles and schedules.

Figure 6.6 Boston Cambridge Library

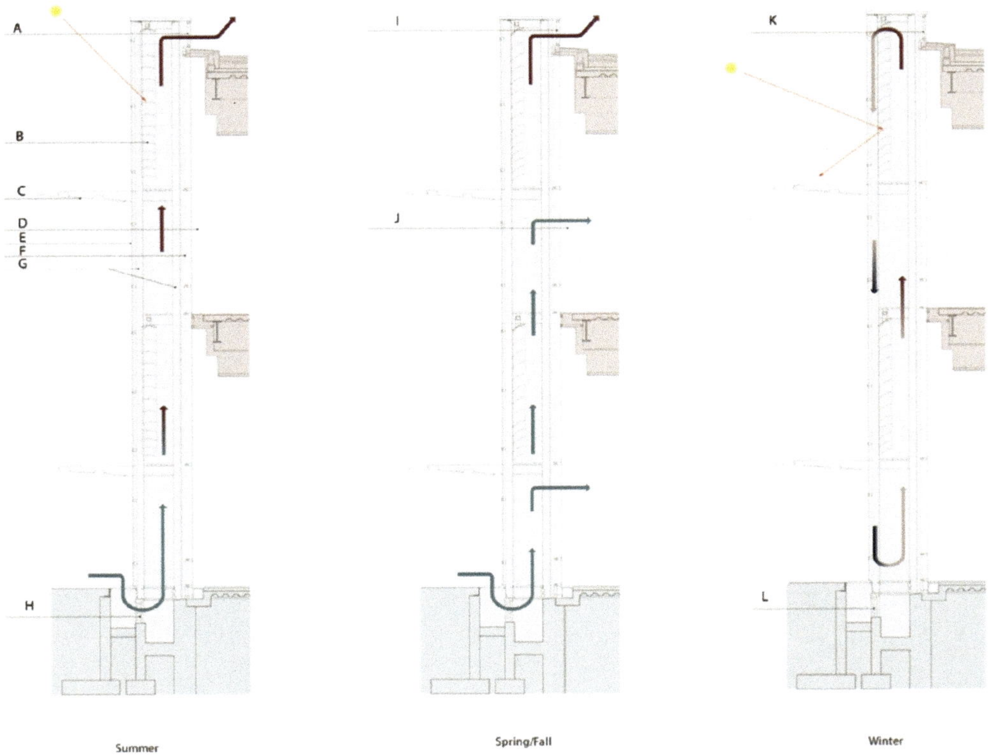

A Exhaust vent with operable glass louvers
B Operable Aluminium louver
C Glass visors
D Operable windows
E Outer skin
F Inner skin
G Structural frame
H Intake trench with operable dampers

I Operable louvers can be opened/closed to control the temperature inside the cavity
J Operable windows for natural ventilation during shoulder seasons
K Operable louvers closed in winter
L Operable dampers closed in winter

Figure 6.7 Section Drawings – Performance in Different Seasons

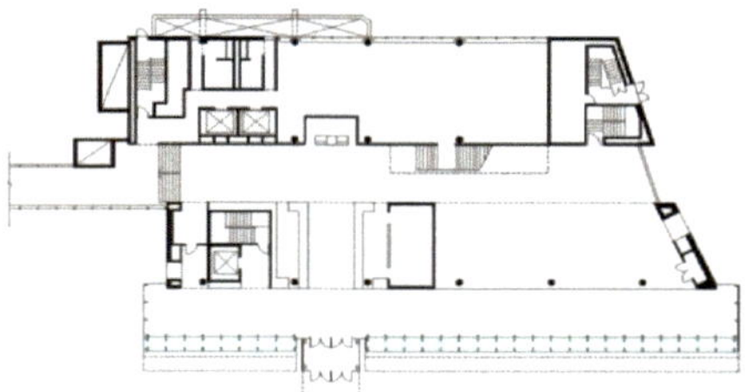

Level 1

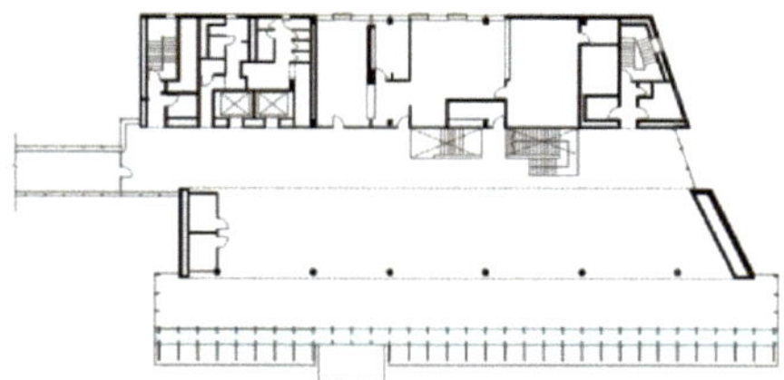

Level 2

Figure 6.8 Plan Drawings

These schedules are often repeated floor to floor. The atrium is treated as one single zone inside DesignBuilder, with openings in the floor and ceiling of the corresponding floors connecting them.

After obtaining basic building and weather data, a simulation of the building's energy performance was undertaken. This was the base case simulation for a single skin façade system. The simulation software used for this research was DesignBuilder, with EnergyPlus as an engine. Since the behavior of double skin façades is highly dependent on the type of climate, the following sections outline a process for selecting the type, analyzing characteristics and properties and selecting strategies preferable for a specific context.

Components of the studied double skin included an internal layer, composed of 6 mm clear glass pane, 20 mm argon filled gap and 6 mm internal low-E glass and aluminum frame with thermal breaks. The external layer consisted of 6 mm clear glass.

Components of the single skin were similar, with double glazing composed of 6 mm external clear glass, 20 mm argon filled gap, internal 6 mm low-E glass and aluminum thermal break frame. Both cases included shading blinds, where in the case of double skin, they were placed in the cavity, and for double glazing on the interior of the building.

BUILDING PERFORMANCE ANALYSIS

Monthly energy results average the daily variation and show how energy loads respond to seasonal variations. It can be noted that lighting loads are reduced during the summer, as the days are longer and daylighting can meet a larger portion of the lighting needs. There is also a reduced fan load during the summer months. However, these savings are offset by a roughly 40% increase in cooling loads during the summer months and the total electric load remains fairly constant throughout the year. The benefit of the double façade becomes apparent when comparing heating loads. The annual energy consumption of the double skin has been compared with single skin as

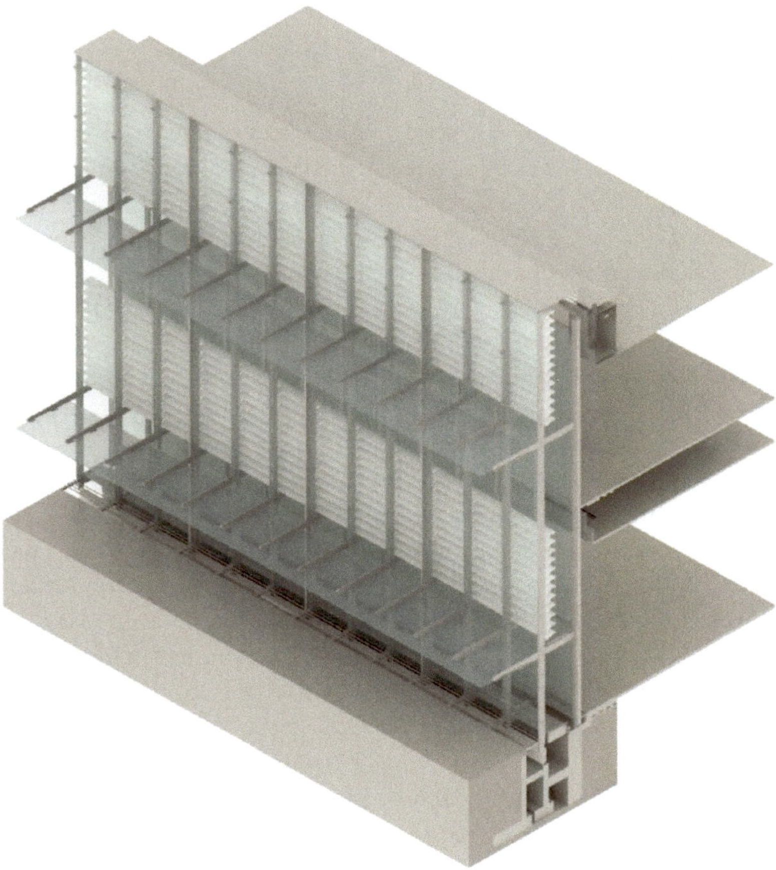

Figure 6.9 Detail and 3D Sections of the Façade

Figure 6.10 View to the Double Skin Façade

illustrated below. It was discovered that in total, the double skin façade consumes 5% less energy than a single skin façade (Figure 6.12).

MODIFICATIONS TO THE BASE CASE

In the second part of the analysis, the double skin façade system has been compared with four different alternatives illustrated in figure 13. The cavity size has been changed from 3 ft to 5 ft. The U-value of the internal glazing and external glazing has been improved from .44 to 0.246 for internal and from 0.56 to 0.449 for external glazing respectively. The results of these alternatives have been tabulated and illustrated.

CONCLUSION

Learning experience through Energy simulation and Natural ventilation analysis using DesignBuilder is applicable in the real world, where you want to check the building performance in a design phase. The comparison of the actual consumption of a building with the DesignBuilder base case proves the capabilities of the simulation tool.

One of the limitations of this tool is that the data available for the building functions is general and not specific to different types of building activity. The other limitation of the tool is that after many changes and many simulations in that same model, many times it does not work for CFD analysis.

Double Skin Facade

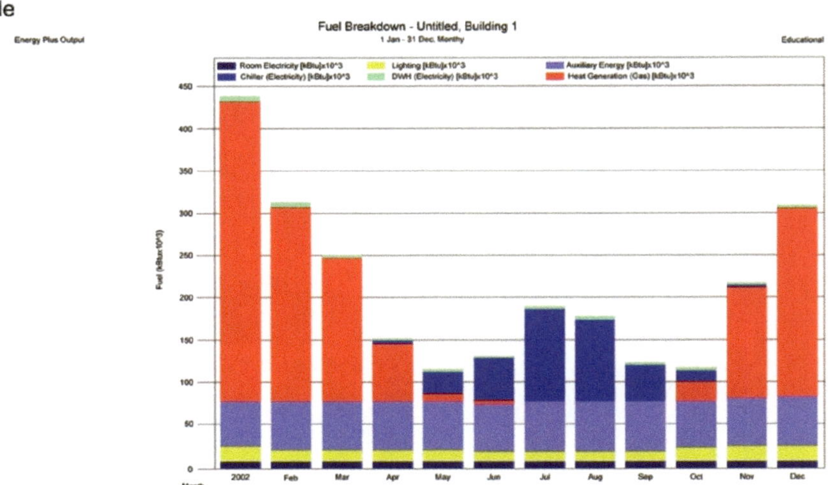

Month	2002	Feb	Mar	Apr	May	Jun	Jul	Aug	Sep	Oct	Nov	Dec
Room Electricity [kBtu]x10^3	7.62	6.73	7.23	7.32	7.62	6.94	7.62	7.43	7.13	7.62	7.13	7.43
Lighting [kBtu]x10^3	17.48	13.65	13.63	11.89	11.54	10.69	11.25	11.28	11.81	14.52	16.07	17.70
Auxiliary Energy [kBtu]x10^3	53.98	59.75	57.82	59.75	57.82	59.75	59.75	57.82	59.75	57.82	59.75	59.75
Heat Generation (Gas) [kBtu]x10^3	355.22	230.33	169.93	69.58	10.69	1.31	0.00	0.00	0.16	23.46	132.05	221.40
Chiller (Electricity) [kBtu]x10^3	0.00	0.00	0.75	1.16	25.64	51.11	109.44	99.51	43.49	12.00	0.96	0.04
DWH (Electricity) [kBtu]x10^3	2.15	1.88	1.98	2.06	2.15	1.89	2.15	2.07	1.98	2.15	1.98	2.07

Single Skin Facade

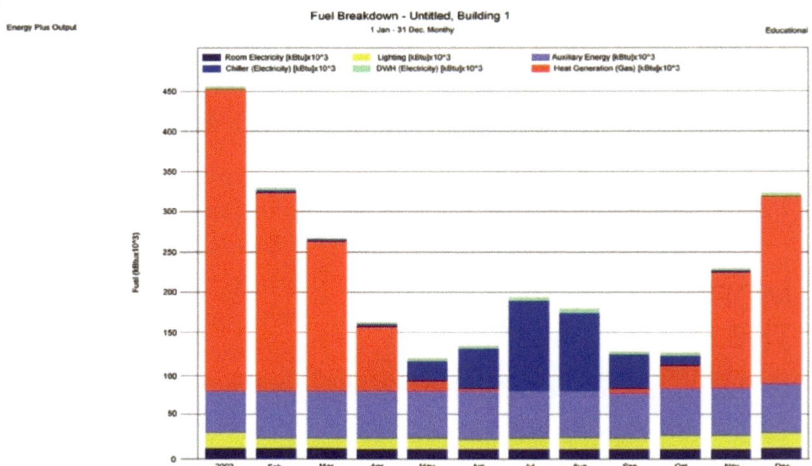

Month	2002	Feb	Mar	Apr	May	Jun	Jul	Aug	Sep	Oct	Nov	Dec
Room Electricity [kBtu]x10^3	10.78	9.50	10.16	10.35	10.78	9.73	10.78	10.47	10.04	10.78	10.04	10.47
Lighting [kBtu]x10^3	17.92	13.93	13.68	11.89	11.57	10.72	11.28	11.39	12.05	14.84	16.51	18.24
Auxiliary Energy [kBtu]x10^3	54.35	60.17	58.23	60.17	58.23	60.17	60.17	58.23	60.17	58.23	60.17	60.17
Heat Generation (Gas) [kBtu]x10^3	370.46	242.14	180.83	76.99	12.57	1.90	0.07	0.07	0.77	27.63	139.85	232.20
Chiller (Electricity) [kBtu]x10^3	0.04	0.22	1.10	1.52	25.33	60.58	106.59	98.04	42.14	12.89	1.86	0.32
DWH (Electricity) [kBtu]x10^3	2.29	2.00	2.11	2.19	2.29	2.01	2.29	2.20	2.10	2.29	2.10	2.20

Figure 6.11 Energy Simulation of DSF and Single Skin Façade

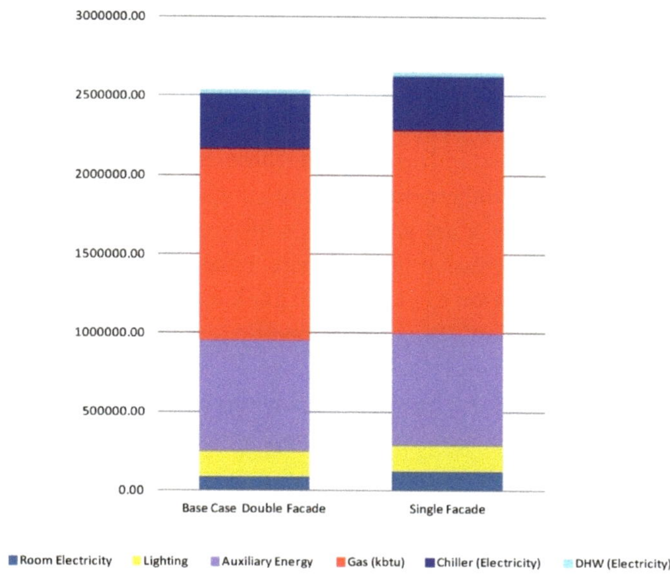

Figure 6.12 Energy Simulation Comparison

However, making a model geometry and building specification inputs is not very time-consuming if you know the tool very well. From the climate analysis, it is inferred that building orientation has been decided considering the prevailing winds; which can make the natural ventilation possible in the summer and/or in shoulder seasons. Here CFD analysis through DesignBuilder shows the temperature, wind velocity and comfort contours. CFD analysis takes a lot of time though it is not very accurate. The CFD analysis was done for April, July and September does not show the air intake and air outtake properly as it should work in the real building. But the temperature, PMV [Predicted Mean Vote on the thermal sensation scale], PPD [Predicted Percent of Dissatisfied people at each PMV] contours are helpful to understand human comfort affected by the double skin façade in the building.

In general, a real-life case-study helps to understand the effect of building location, orientation, building activity, building operating schedule, HVAC, lighting and equipment properties, building materials; together with the building energy performance prediction.

INTERVIEW

When you designed this building, did you consider another DSF type?

In a search for precedents, we examined many different existing projects located here in the United States, and participated in a research trip to Germany and England. While visiting Europe, we discovered double skin façades (DSFs) utilized in twelve structures that could be studied further and used for inspiration. In the United States, we traveled to the Manulife and Genzyme facilities located in Boston, MA. The Genzyme facility incorporates a thin system inside the building, but requires a high amount of maintenance.

With regard to the system used at the Genzyme facility, all windows in the entire building must be opened individually for cleaning purposes. If implemented in our building, cleaning the glass windows would involve an intensive and difficult process, as the floor-to-floor height is taller than that of the Genzyme facility. Both a system of ladders and several separate openings at altering heights would be required.

In Europe, we viewed many other existing systems of DSFs, some of which inspired our current system. The system that we studied includes individual fins about eight inches deep. Another involves corridors that bend at the bottom and the top of each individual floor level. Others are utilized at the full building height, i.e. multi-story.

To further comprehend which system would be best for our application, we hired Arup Façade Group as a consultant and worked very closely with them. They helped us evaluate the right form for the façade and the most conducive type of technology for our building. This involved learning that techniques we may have found innovative and effective, might not be as effective if implemented for our specific use. For example, in Germany, we examined designs that were incredible in terms of sophistication. This included a few of Murphy Jahn's projects, such as the post tower. Some structures utilized robots that moved back and forth through the double skin, cleaning the glass as they moved. Others contained sunshades that opened and closed based on individual sun sensors. With sun sensors in use, louvers respond to sunlight and close automatically when the sun's rays are most powerful. Although many of these techniques are innovative and inspiring, such complex systems could not be implemented in our project. We were financially limited and only used technological systems sparingly and with specific intent.

How does the system work and discuss the dimensions?

We aimed to use a system of thermodynamics with the ability to vent passively. We did not want to utilize any active fans to direct air through the deep cavity, high flute, and building glue. Instead, we aimed to create a stack effect that would keep the unwanted heat away from the building through passive design.

In our DSF, the cavity is about three feet deep and the building is three stories in height. The first two floors contain relatively high floor-floor heights. The first-floor height reaches 19 feet floor-to-floor and the second-floor amounts to about 17 feet. In comparison, the third floor sits back 20 feet from the lower two levels, and is used for children's activities. This setback is not due to the massing requirements that the neighbors imposed on the building, but instead is a response to the height of the existing building. Therefore, the third floor is set back from the existing front façade.

In an attempt to break the mass of the building down visually and assign a datum line that responds to a singular point in the children's room, the children's floor was designed to be transparent, with views of the park. As a result of the transparent design choice, the seating areas are located in the middle of the building on each floor, not on the perimeter. In this way, all floors resemble a treehouse and contain a variety of seating areas. In this way, the façade allows for a reduction in the amount of glass used on the children's floor, and ultimately a more transparent view outside the building.

How does the façade operate and ventilate?

The façade reacts to exterior heat temperatures differently during each season. During the winter months, the façade will shut and allow heat to build up in the cavity. If the heat builds up to 80 degrees in the cavity, the vents at the top and bottom will open to exhaust excess heat. While most common in winter, this will occur during any season if overheating occurs. During spring and fall, the manual windows can be opened from the interior of the building. This works as a passive heat recovery strategy that moves heat from the cavity to the interior of the building. In the summer months, the windows can remain closed to stop heat from entering the building.

With this technique, opening the windows in the wintertime may not transmit heat out of the façade if the sun is not shining. Instead, cold air will be cycled into the building. Nonetheless, there are certain periods of time in the winter months when windows can be opened and the temperature of the cavity will be higher than inside the structure. This technique can be used to reduce heat loads.

Did you harvest any heat from the cavity?

In the winter months, the façade provides a thermal buffer between the exterior weather conditions and the interior spaces. Drawing away small amounts of heat can be beneficial. However, if too much heat is drawn away from the building, the principal benefit for our project, comfort, is reduced. The mechanical room for all systems of the building is located in the basement. If heat were to be transferred through air handling units throughout the building, air would be required to be pushed through ducts with the introduction of fans controlled from the basement. This technique is not as efficient as capturing heat in the façade, and thus the building relies mainly on the passive techniques of the façade for heating and cooling.

In the summer months, the main goal is to keep direct sun rays from entering the building. This could be accomplished by the use of a single skin curtain wall with external sunshades. However, these sunshades would be large, heavy, and expensive if not placed in the cavity. In contrast, we placed these shades inside the cavity. This allowed us to use relatively light and inexpensive screens. The DSF will work as a barrier to protect these screens from dirt, potential damage, and exterior weather conditions. The sunshades have the potential to be automated and control low sun angles from entering the interior of the structure. In addition to sunshades, blinds can be placed outside of the building's thermal barrier, instead of the inner skin, so that heat gain does not enter the building. To ensure that these shades are utilized correctly, the façade must be vented to prevent the cavity from building up excess heat.

How do the sunshades operate?
The shades rely on a simple time clock. We aimed to keep this system as simple as possible to avoid possible maintenance issues for the library space. Many systems we studied in Europe relied on sun sensors. With this technique, if the sun is covered by a cloud at any point in the day, the shades react and contract accordingly. Although very efficient with regard to solar gain, rotating screens that contract up to10 to 15 times a day have a high potential to break. The possibility of the motor breaking and requiring maintenance is a big concern in our application. Therefore, our motors are relatively simple and can be compared to motors controlling an internal shade in a conference room. The more the motors are used, the more they are subject to breaking. Our motors will remain horizontal in the winter and close around 12 to one o'clock in the summertime.

Do you take advantage of redirecting the lights with these shading devices?
Yes, daylight is naturally reflected due to the louvers' shaped and curved design, which helps bounce light to the desired angle. In addition, the ceiling is painted white in the first zone of the structure to ensure that light reflects off the ceiling and deeper into the building. Behind the white aluminum ceiling is a wood ceiling, and further in the middle portion of the building is another aluminum ceiling, all strategically placed with light in mind.

What was the intention behind having the louvers instead of shades that come all the way down? You have one canopy on the first floor.
Many of the projects we viewed in Europe utilized adjustable louvers that blocked any views out of the building. If the sun is shining, the only view out of the building is through a louver. In our application, we used automated shades to control the light from eight feet above the floor. We then placed an external glass visor to block the sun from entering the lower eight feet of each floor. Therefore, there is no louver blocking views out of the structure. The glass sunshade is additionally placed on the outside of the structure, creating a DSF that is unique to any other application in the world.

Can you talk about the construction and structure of the façade?
The glass visor is the single most complicated structural aspect involved in our project. It cantilevers about six feet from the façade of the building and is located eight feet above the floor level, on a 20-foot floor to floor height. The visor ladder trusses run in front of the inner and outer skin system and support the cantilever. This truss system does not involve any diagonals and was fabricated in Germany by Gartner before being fabricated individually. The inner skin of the façade is connected to the ladder truss at three or four points in between the floor levels. The ladder truss itself acts as the structure. The inner skin is attached to the ladder truss. The ladder truss is then attached back laterally only on the second floor, where it is supported

by the floor. The first floor contains a concrete base with a ladder truss which contains outriggers for the visor. The ladder truss contains a lateral connection on the second floor and is also supported by the floor. This is done so that there is no gravity load that can tie back to the floor. Because these floor levels are cantilevered, there are no columns in use. The building's structure cantilevers 8 feet from the inner skin which then comes in and is connected back at multiple points. As a result, the outer skin is attached with patch connections to the trust.

With the methods discussed above, the mullions can remain relatively thin on the inner skin. No mullions are present on the outer skin. The outer monolithic glass contains patch fittings that attach back to the ladder truss system.

How does it perform?

The projection shows that the double skin is 50% more efficient than a single skin system. This prediction is not depicting that the building will save 50% of total energy, because it is only applied to one of the four façades. Rather, this prediction indicates that if the south façade were to be changed from a double skin curtain wall to a single skin curtain, all of the heating and cooling loads that would be generated would be 50% higher at that point, relative to the surface area of the glass itself. To summarize, the façade itself would be performing 50% less efficiently if it were a single skin façade as opposed to a double skin façade.

The entire façade was fabricated in Germany and is placed at about five and a half feet wide by 12 feet tall.

Bibliography

Azarbayjani, M. (2011). Integration of natural ventilation in a building with double skin façade. BESS 2011, Pomona, CA.

Cambridge Public Library, Cambridge, MA, Case Study: A Double-Skin Glass Wall, published report by William Rawn Associates, Architects, Inc. with Ann Beha Architects. Retrieved from: https://rawnarch.com/pdf/CPL_WRA.pdf

Navvab, M. (2005). Full scale testing and computer simulation of a double skin façade building. 9th International IBPSA Conference Building Simulation 2005, Vol 2, Montreal (Canada). 831–838.

Safer, N., Woloszyn M., Roux J.J., Kuznik F. (2005). Modeling of the double skin facades for building energy simulation: Radiative and convective heat transfers. 9th International IBPSA Conference Building Simulation 2005, Vol 2, Montreal (Canada). 1067–1074.

Stec, W., and Passen D. (2003). Defining the performance of the double skin façade with the use of the simulation model. 8th International IBPSA Conference Building Simulation 2003, Vol 2, Eindhoven (Netherlands). 1243–1250.

Torres, M., Alavedra P., Guzman A. (2007). Double skin facades – cavity and exterior openings dimensions for saving energy on Mediterranean climate. IBPSA Proceedings of Building Simulation 2007, 198–205.

Figure 6.13 John E. Jaqua Academic Center

John E. Jaqua Center for Student Athletes

LOCATION
Eugene, Oregon

BUILDING PROGRAM
Multi-Use

ARCHITECT
ZGF Architects LLP

SIZE
37,000 SF

COMPLETION DATE
2010

BACKGROUND AND CONTEXT

The John E. Jaqua Academic Center for Student Athletes was established in 2010 for students' athletes of the University of Oregon in the city of Eugene. The building is surrounded by a scenic site, including landscape and a reflective pond with a constant flow of water, which creates an encouraging environment for learning. The 37,000 SF John E. Jaqua Center consists of a variety of functions and spaces within three different levels in order to respond to student athletes' needs. The first level is open to the public and includes a cafe, auditorium and an atrium for events.

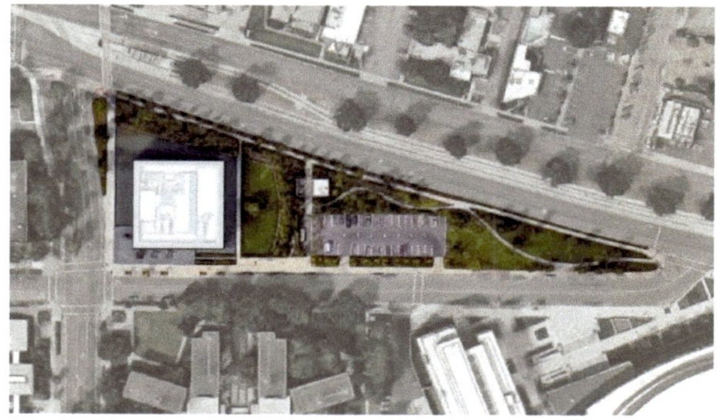

Figure 6.14 Building on the Site

Eugene, Oregon has very mild and temperate weather in comparison to other cities and states. The temperature fluctuates between about 5 degrees Celsius in the winter months and over 25 degrees

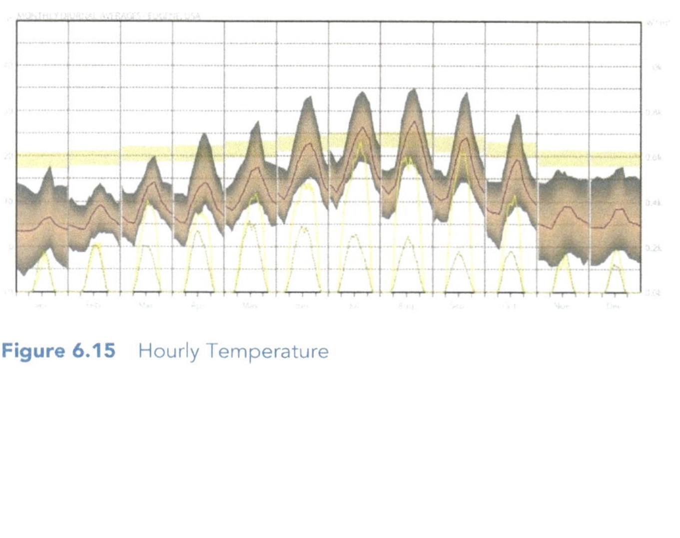

Figure 6.15 Hourly Temperature

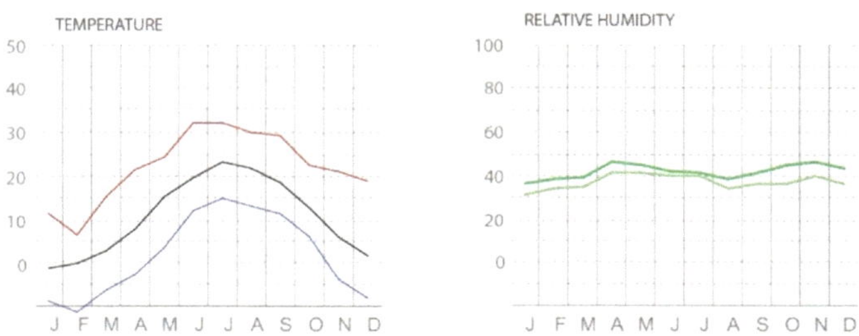

Figure 6.16 Climatic Data – Monthly Temperature, Relative Humidity

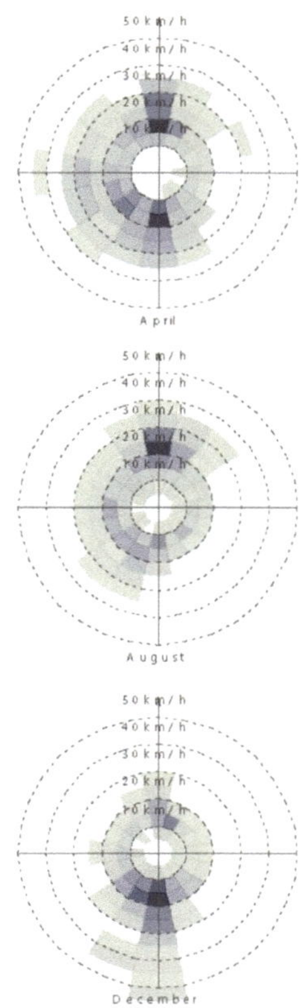

Figure 6.17 Prevailing Winds

Celsius in the summer months. This can often cause issues with regard to the heating and cooling of the building, as it is difficult to maintain a stable internal temperature. Seven months of the year, the average temperature in Eugene remains around the personal comfort zone. Furthermore, because of the excessive number of overcast days, the direct solar heat gain received by the building is less than normal, which is also offset by the double skin façade system and the shading devices in the cavity.

DESIGN INTENT AND BUILDING FEATURES

The Jaqua Center has a variety of exterior systems to help control its internal temperature as well as the lighting on the inside. The most notable piece is the double façade on the exterior of the building. This façade wraps all the way around the building, and does not transfer air directly into the building. There are also vents at the very bottom of the façade system, and at the inward top part of the façade. The vents are operable and are set to a schedule which allows for air movement when the heat and additional piece on this building is the skylight. This skylight puts light into the atrium, which is open to all the floors for light distribution into the interior spaces. This contributes to the lower lighting load during the daytime hours. Floor-to-ceiling vision glass allows natural daylight to flood the interior space and a stainless-steel screen forms a layer within the glass cavity. Floor-to-ceiling draper solar shades were motorized with 138 LT50 ILT 2 motors, installed on the exterior of the inner glass façade. The sunshades are controlled via the Somfy Digital Network (SDN) system. Externally, the stainless-steel screen provides privacy for users; internally, there is complete visual transparency with views over the landscape and gardens. The double skin wraps a blanket of warm air around the building and mitigates the highs and lows of temperature fluctuations. This insulating façade also provides superior acoustics, helping create a conducive atmosphere for study and learning. LED lamps at the base of the outside wall illuminate the exterior at night, turning the skin into a translucent glowing box. "

The specifications of the project are: A Unique High-Performance Façade:

- The double skin façade (two surfaces of glass, creating insulated airspace) is a multi-story (full height), full depth (3'), thermal flow. The façade allows for complete transparency while ensuring protection from excessive heat gain, heat loss, and glare.

Energy Savings and Comfort

- The cavity can be open in summer to keep heat from entering the building and closed in the winter to create an insulating "thermal blanket."

Natural Light

- The façade brings a significant amount of balanced natural light into the building, carefully controlled by fixed and movable sunshades.

Figure 6.18 Cavity

Figure 6.19 Detail

Figure 6.20 Lobby

Figure 6.21 Atrium Space

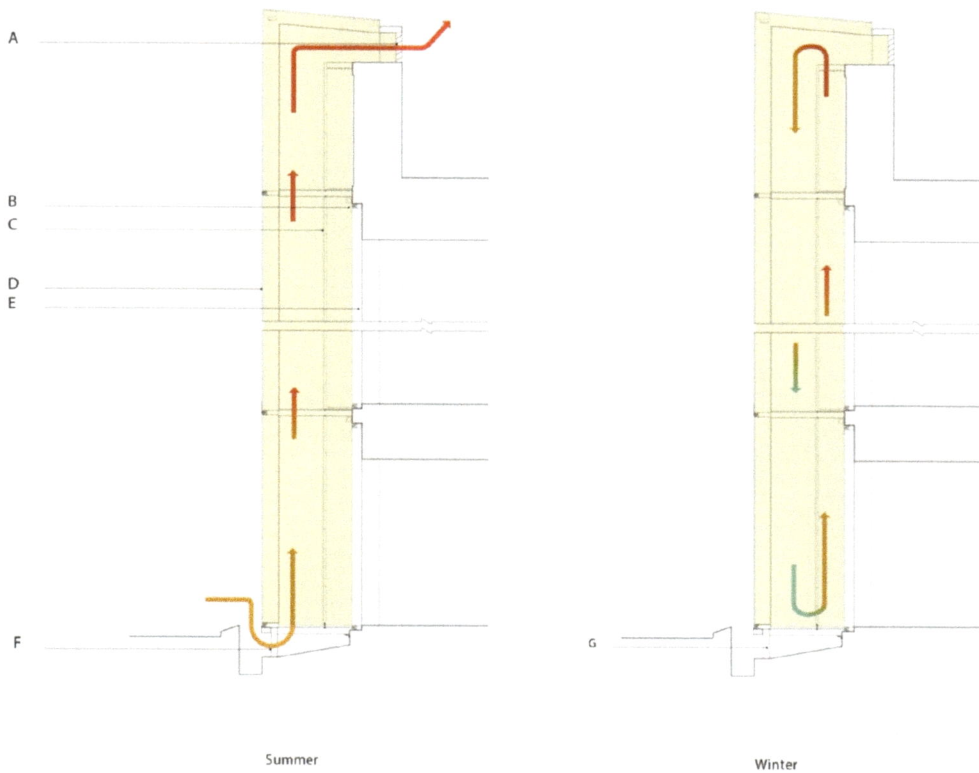

A Operable louvers / opened in summer
B Adjustable blind
C Fixed shading device
D Outer skin / mullion system
E Inner skin / mullion system
F Intake trench with operable dampers

G Intake trench with operable dampers

Figure 6.22 Section Drawings, Performance in Summer and Winter

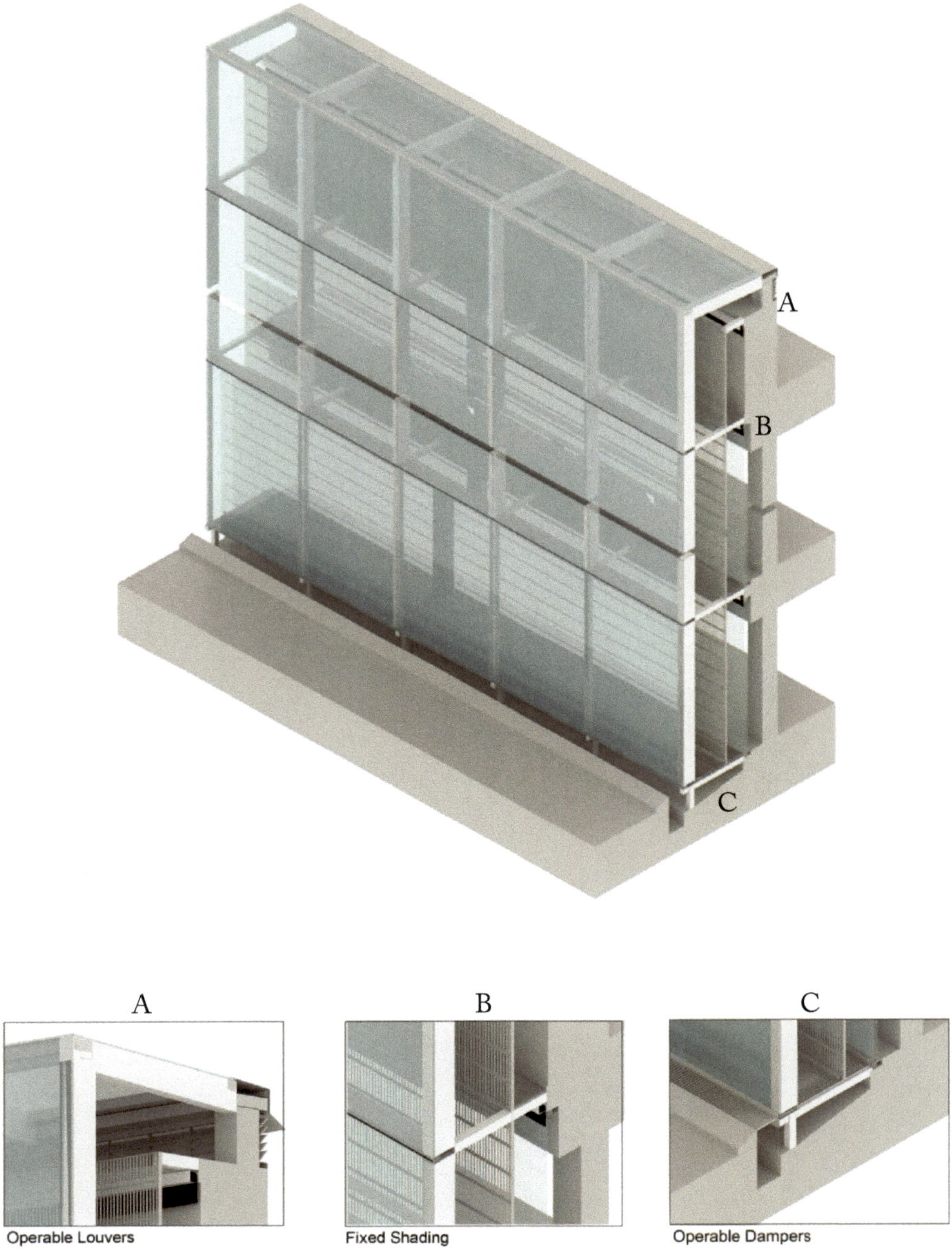

Figure 6.23 3D Section and Details (a) Operable Louvers (b) Fixed Shading Device (c) Operable Dampers

Natural Ventilation

- The double skin façade ventilates to the outside during the shoulder seasons.

Floor-to-ceiling vision glass allows natural daylight to flood the interior spaces and a stainless-steel screen forms a layer within the glass cavity.

- The façade is attached to the building slabs like any other curtain wall.

BUILDING PERFORMANCE ANALYSIS

The monthly energy results average the daily variation and show how energy loads respond to seasonal variation. It can be seen that lighting loads are reduced during the summer, as the days are longer and daylighting can meet a larger portion of the lighting needs. There is also a reduced fan load during the summer months. However, these savings are offset by a roughly 40% increase in cooling loads during the summer months and the total electric load remains fairly constant throughout the year. While the values do not correspond with the measured data, level electric load through the year supports our findings. There is some variation through the year (plus or minus 10% with a minimum in the spring), however it is unclear if this is statistically significant.

The plan was divided into multiple zones according to geometry, proximity and use. Each zone was then assigned an activity template (indicated by color) which associates that zone with different energy profiles and schedules. These schedules are often repeated floor to floor. The atrium is treated as one single zone

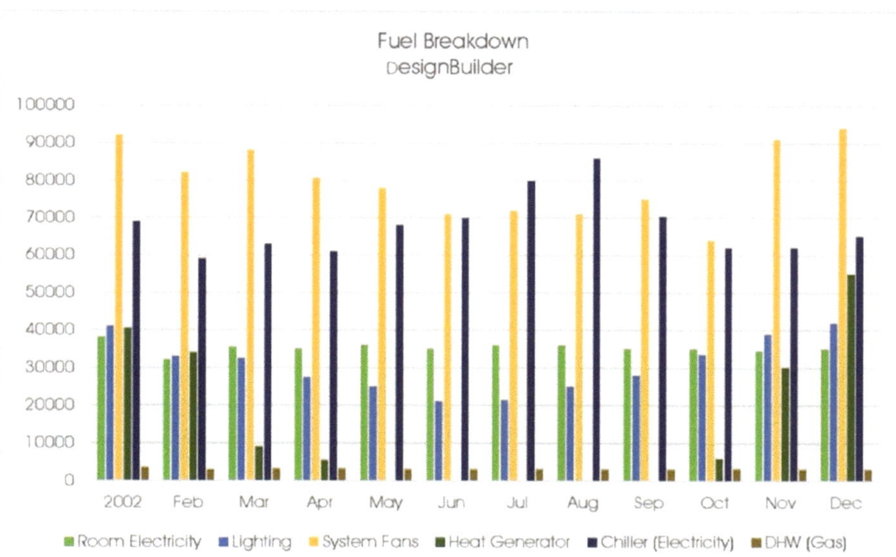

Figure 6.24 Fuel Breakdown

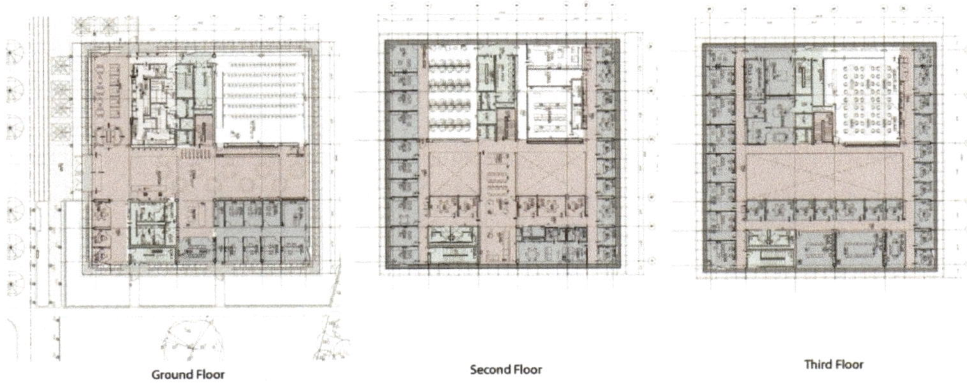

Figure 6.25 Plan Drawings

inside DesignBuilder, with openings in the floor and ceiling of the corresponding floors connecting them.

The default schedules for the library activity templates were based on a 9 hr/day, 6 days/week operational schedule. Outside this time, plug, lighting, heating, cooling and DHW loads would shut off. However, being a collegiate facility, the Jaqua Center maintains significantly longer hours. For this model, the schedule was simplified to a 14hr/day (8 am–10 pm) 7 days/week. While this schedule does not reflect daily variation in library hours, it averages the same number of operational hours per week. The CFD analysis for the summer afternoon suggests that the double façade is performing as expected. Cool air is entering the bottom of the façade cavity and warm air is exiting through the top. However, it would appear that the CFD editor has assumed that the top of the third floor is at the top of the façade, which, in actuality, extends above the roof to form a kind of parapet. This would appear to be causing the façade to vent into the third floor in the simulation and the third floor and atrium are overheating.

The model was used to compare the different glazing types for the interior layer of the double façade and to compare to a similar building with an all-glass single façade. It was found that

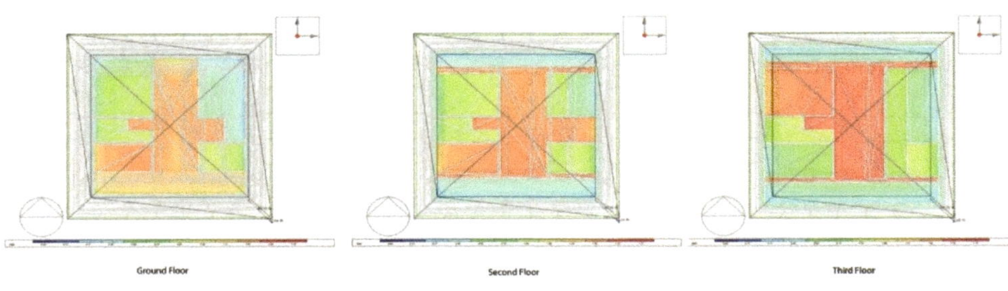

Figure 6.26 Jaqua Plans – Predicted Mean Vote, Comfort Analysis

Velocity	0.2	10.63	20.37	31.30	41.54	51.97	62.21	72.64	82.88	93.31	103.55	113.98	(ft/min)
Temperature	60.04	60.06	61.68	62.50	63.32	64.13	64.95	65.77	66.59	67.41	68	69.04	(F)
PMV	-0.06	0.11	0.27	0.43	0.60	0.76	0.93	1.09	1.25	1.42	1.58	1.75	

Figure 6.27 Temperature Stratification from First to the Top Floor

Velocity	0.2	10.63	20.87	31.30	41.54	51.97	62.21	72.64	82.88	93.31	103.55	113.98	(ft/min)
Temperature	60.04	60.06	61.68	62.50	63.32	64.13	64.95	65.77	66.59	67.41	68	69.04	(F)

Figure 6.28 Temperature and Velocity in Double Skin Façade Cavity

a low-E single panel interior glazing resulted in a 12% increase in electric use whereas adding argon to the glazing unit decreased electric consumption by 7%. The single façade resulted in a 10% increase in electric consumption resulting from a decrease in lighting loads and increase in cooling loads as the second façade no longer shaded the building.

The benefit of the double façade becomes apparent when comparing heating loads. Removing the second façade resulted in a 223% increase in heating loads. While the saving of 471,940 kbtu of gas is significant, it should be noted that heating is a relatively small load for a building of this size and heating processes can be done efficiently especially with respect to carbon; such that the carbon impact of this savings is equivalent to a savings of 165,984 kBtu of electricity or 6% of the annual electric load (assuming 0.381#CO2/kbtu for electric and 0.134#CO2/kbtu for gas). Changing the inner glazing to a single pane resulted in a 34% increase in heating loads, while using argon to insulate the inner glazing saved 6%.

Calculating the EUI allows a metric to compare various strategies, which includes both electric and gas loads and normalizes the results to the size of the building so they can be compared to other buildings or national averages. The measured EUI was 114 kbtu/sqft or about 20% less than the CBEC average for comparable buildings.

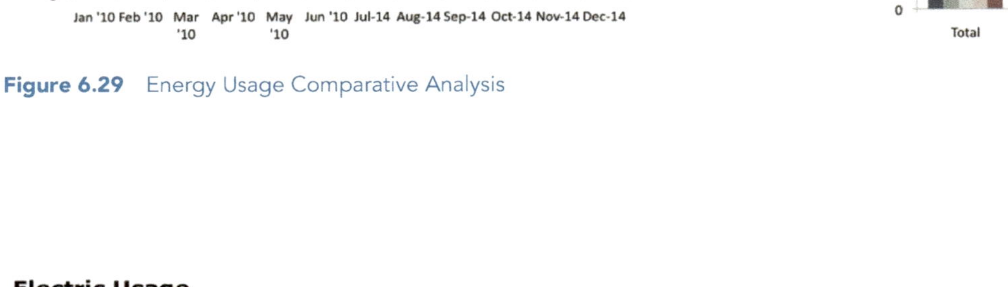

Figure 6.29 Energy Usage Comparative Analysis

Electric Usage

			Jan '10	Feb '10	Mar '10	Apr '10	May '10	Jun '10	Jul-14
Measured Electric		*Kbtu*	*343,513*	*269,886*	*337,832*	*333,642*	*273,878*	*338,375*	*348,713*
Single LowE		Kbtu	267,309	235,051	246,593	228,142	230,771	221,801	236,152
double LowE		Kbtu	238,613	207,004	219,646	202,930	204,615	197,349	210,197
double LowE + Argon		Kbtu	219,930	191,216	203,820	188,154	189,539	181,495	192,754
Singel Façade		Kbtu	258,045	228,261	243,548	226,350	230,572	220,475	229,958

Electric Usage

			Aug-14	Sep-14	Oct-14	Nov-14	Dec-14	Total
Measured Electric		*Kbtu*	*353,251*	*270,759*	*292,535*	*350,525*	*323,649*	***3,836,559***
Single LowE		Kbtu	245,270	233,111	241,639	255,710	270,003	**2,911,553**
double LowE		Kbtu	218,062	207,410	215,748	227,661	238,378	**2,587,615**
double LowE + Argon		Kbtu	200,585	191,429	199,649	209,432	218,993	**2,386,996**
Singel Façade		Kbtu	239,205	229,636	240,217	246,949	259,772	**2,852,988**

Figure 6.30 Electric Usage Tables

Figure 6.31 View to the Building

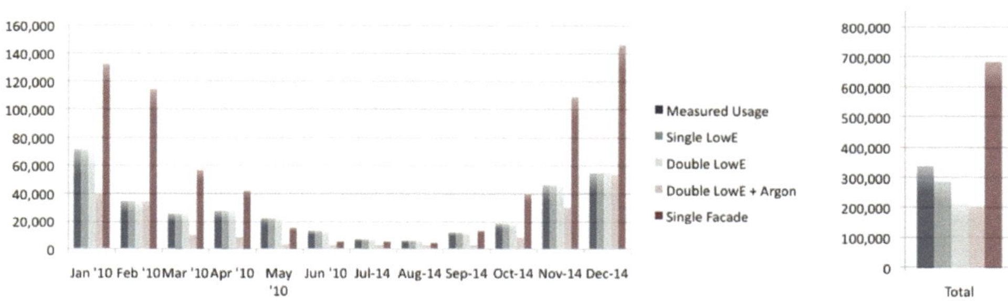

Figure 6.32 Gas Usage Comparative Analysis

Gas Usage

		Jan '10	Feb '10	Mar '10	Apr '10	May '10	Jun '10	Jul-14
Measured Gas	*Kbtu*	*71,010*	*34,280*	*25,200*	*27,510*	*21,930*	*13,180*	*6,890*
Single LowE	Kbtu	59,245	46,940	15,289	11,592	4,104	2,681	2,743
double LowE	Kbtu	43,158	35,896	10,659	8,591	3,500	2,677	2,699
double LowE + Argon	Kbtu	39,788	34,645	9,972	8,190	3,436	2,712	2,702
Singel Façade	Kbtu	132,265	114,339	56,456	41,675	15,759	5,549	5,200

Gas Usage

		Aug-14	Sep-14	Oct-14	Nov-14	Dec-14	Total
Measured Gas	*Kbtu*	*5,800*	*11,770*	*17,750*	*46,010*	*54,520*	***335,850***
Single LowE	Kbtu	2,773	4,272	12,794	46,417	76,465	**285,315**
double LowE	Kbtu	2,677	2,998	8,697	32,619	57,477	**211,649**
double LowE + Argon	Kbtu	2,687	2,946	7,871	29,940	53,544	**198,433**
Singel Façade	Kbtu	4,493	12,639	39,852	108,670	146,692	**683,589**

Figure 6.33 Gas Usage Tables

Figure 6.34 Interior Views (a) to the Atrium (b) Front Door

CONCLUSION

Relative results suggest: Double façade significantly improves energy performance (10% reduction in electric and 69% reduction in Gas)

- Insulated (Low-E double pane vs Low-E single pane) interior glazing results in significant improvement (11% reduction in electric, 26% reduction in gas)
- Argon insulated, interior glazing results in a small improvement over the air (7% reduction in electric, 6% reduction in gas).

While our investigation strongly suggests that a double façade in Eugene's climate will lower EIU, it is less clear that it will result in lower energy costs and associated environmental impacts.

It is suspected that, while a double façade provides a more efficient all-glass box, other architectural solutions could prove more efficient and less costly. It is suspected that less compact massing would allow for more heat loss, lessening cooling loads enough to offset the heating liability such a solution would pose. Exterior shading devices on a single façade could also provide the same shading and day-lighting benefits of the double façade without the heat gain potential of the double façade. While the humid climate might pose certain issues for natural ventilation, it is believed that there may have been missed opportunities to ventilate portions of the building during portions of the year. One could imagine a similar organization where the atrium and circulation spaces could become more of a covered courtyard, which is naturally ventilated. These alternative solutions suggest fundamentally different architectural responses, not just a technical solution, and so it is the conclusion of this investigation that while the double façade of the Jaqua Center is an appropriate and successful solution to the all-glass box in Eugene, it has not been established that the glass box is an energy-appropriate solution for this specific climate.

INTERVIEW

How did you decide to do the double skin for the Jaqua center in the first place?
Our main design motivation behind the use of the double skin façade was the desire to create a unique interior atmosphere with the central component of the building encapsulated in its own "box." We aimed to represent the idea of a small jewel in a garden. The building functions as a study center and library. It is meant to isolate the inhabitants from all surroundings, allowing for studying to be the primary focus.

However, the project is located adjacent to a fairly busy intersection and thus, external noise is an issue that had to be addressed. In order to address exterior noise, interior acoustics must first be studied and addressed. We asked ourselves questions such as: "How do we create an environment inwardly focused on studies?" This question prompted a discussion with the client about the perception of the library space. We asked the client if they would prefer a library that is completely enclosed with no visual connection to the outside or if the space should be similar to a classical Chinese Garden, where surrounding nature is beneficial to studying and provides a peaceful atmosphere that boosts creativity. A transparent box was the result of these conversations.

Our goal is to allow those in the library to appreciate the landscape that surrounds it, but at the same time, not be distracted by external factors. To implement this concept, the building is almost entirely glass on the exterior, with the exception of one staircase that contains insulation around the back wall. In total, 85% of the skin is glass, but in perceptual terms, 100% of the building appears to be glass. While staying in accordance with the Oregon energy codes, it is rather difficult to design a building that's more than 50% glass. We were required to provide complicated energy models to validate that the structure is compliant with the enforced energy codes. In order to create a building with 85% glass that is also code-compliant, a double façade was necessary to decrease the U-value and solar heat gain coefficient of the envelope.

How does the sun shading work and what are the dimensions of the façade?

The sun shading is a stainless-steel mesh screen created from triangular wires located behind the screen. The outer skin consists of a single-paned system while the inner skin functions as a double screen system with an additional shading device located in the cavity behind the outer screen. The shades follow a time clock, so that the users in the building do not have any responsibility to alter the shading features. This fixed screen is about 45% opaque and is only partially responsible for shading the interior spaces.

In addition to the screen system, automatic shades are located behind the screen in that exterior cavity. As the sun moves around the building, the shades precede the sun by about 40 minutes, reducing any possible glare. The entire façade contracts down the exterior uniformly. In this way, the users of the building can only control the shades by speaking directly with the building maintenance personnel.

We aimed to create an exterior screen system with a uniform appearance. We strived very hard to make no deviations from the module of the "box." To stay in accordance with this idea of a "box" the front doors are five 5' wide and 12'6" tall. The glass panels around the exterior match these dimensions exactly.

The building is three stories tall with a top parapet. On the backside of the parapet are operable dampers which control airflow into the cavity. Each floor is 12'6", with the exception of the top floor, which reaches 10'.

How does it perform during the summer and winter months?

The cavity is open at the bottom, allowing air to be let out on warm days. The building sits above the pond by about 14 inches, which alternatively allows air to enter from the bottom of the cavity as well. This opening is not operable, allowing for air to enter or leave at all times. However, a screen is located at the bottom of the cavity to prevent insects or outdoor elements from entering the cavity while still allowing air transfer. On the backside of the parapets, operable dampers use temperature and pressure sensors to control their movement. When the air and the cavity drop to 55 degrees, this damper closes and holds the air like a warm blanket around the interior.

We recognize that there is a dynamic relationship between the U-value and the solar heat gain coefficient that has the potential to alter this space. However, this depends on whether the damper is up or down and how much heat is available from the sun. We hired an engineering solutions company, Arab Engineers, to do a study of the cavity to determine what the equivalent solar heat gain and U-value would be for a non-dynamic wall and a wall with punctured openings and insulated opaque panels. This proved beneficial for determining terms that the Department of Energy was able to evaluate with their models.

We never initially considered trapping the air inside of the cavity. We assumed that we would let air flow freely on warm days and keep the air in as best as we could on cold days. Our goal was never to hold a large amount of heat in this cavity. However, this skin performs very efficiently in the summer months, while a preferable U-Value of thermal performance is sought after in winter months. Therefore, the building has no separation between the façades, so air moves freely around the entire façade.

We predicted a 30% rise in temperature from the ground floor to the upper floor. In practice, the rise in temperature is much less than predicted and varies throughout the day. The most dramatic temperature change recorded was a 20 degree rise from the ground floor to the upper floor. We suspect this is due to the fact that the air can simply move on its own. Rather than reaching the top of the building and remaining stagnant, air can move freely throughout the façade.

Does the façade perform as expected?
Yes. Specifically, in the winter months, the façade performs well with regards to consumption and usage. The U-value in the cavity is performing as expected. We modeled two different scenarios when researching how best to transfer heat in the cavity. One of our models used warm air from the cavity and fed it back into the heat recovery system. This model concluded that it is more efficient to keep the air in the cavity than to transfer it back into the heat recovery system. Therefore, we have been using this technique. With regards to electricity consumption, there is a pump running throughout the day which generates a large amount of energy, but the façade itself is not using an overwhelming amount of energy.

Describe the daylighting in the space.
Occupants of the building have different desires when it comes to how much light they would like inside their offices. We modeled daylighting for this building around the perimeter and found that 20% of the available daylight was entering the building through this assembly.

What is the construction type?
This building is constructed from a light steel tube structure. The steel members are 2' by 4'. A steel cage is built into the building with the steel elements attached to the slab. The vertical and the horizontal members of the steel are clad with the curtain wall system, aluminum, and unit-sized cassette assemblies that are covering the aluminum. The perception is that the curtain wall system is aluminum; however, it is not.

The inner glass is a slab-to-slab system set on a steel tube that is positioned around the perimeter of the floor. We did not want to break the thermal bridge in place, and instead fabricated a steel collar around the slab and insulated the steel.

The steel tube then extends and is joined with a connecting vertical element. The aluminum essentially "clicks" right onto the face of the steel tube and contains a wrap on the backside. The curtain wall system is aluminum and the screen inside the curtain wall is stainless. The shadings connect to small tabs that are welded onto the back of the panel and bolted in place on two tabs welded onto the steel frame.

Bibliography

John E. Jaqua Academic Center for Student Athletes. Services for Student Athletes · University of Oregon. Web. 05 Feb. 2016. http://ssa.uoregon.edu/

University of Oregon John E. Jaqua Academic Center for Student Athletes / ZGF Architects. *ArchDaily*. 2011. Web. 05 Feb. 2016. http://www.archdaily.com/137141/university-of-oregon-john-e-jaqua-academic-center-for-student-athletes-zgf-architects ISSN 0719–8884.

Figure 6.35 The Richard J. Klarchek Information Commons

THE RICHARD J. KLARCHEK INFORMATION COMMONS

LOCATION
Chicago, Illinois

BUILDING PROGRAM
Digital library [class-rooms, seminar rooms, open areas, commons with computers, cafe]

ARCHITECT
Solomon Cordwell Buenz

SIZE
69,000 SF

COMPLETION DATE
2007

BACKGROUND AND CONTEXT

The four-story, 70,500 square feet Klarcheck Information Commons integrated sustainable design features with innovative mechanical systems to reduce the building's energy consumption by more than 50 percent. Although the initial construction cost for the Information Commons was higher than conventional, energy efficiency measures allow for lower operating costs and a greater long-term rate of return.

Figure 6.36 Building on the Site

The primary axis of the building is oriented nearly north-south with large glass exposures on the east and west sides. The exterior envelope includes a double skin

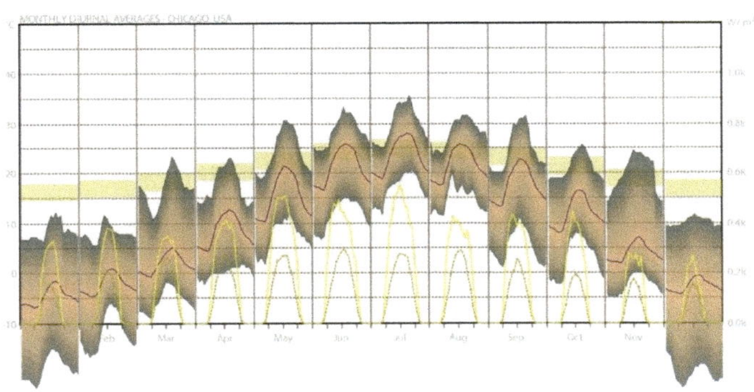

Figure 6.37 Hourly Temperature

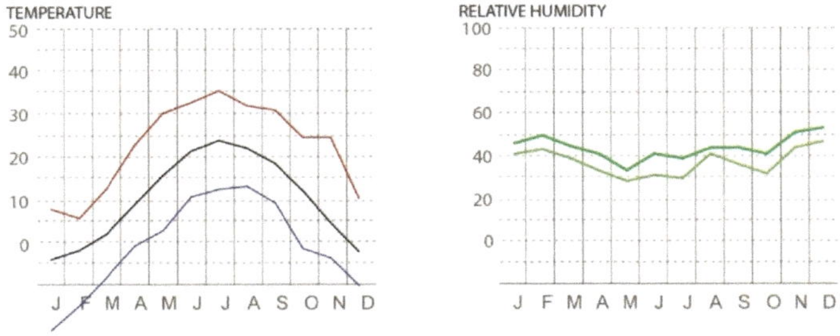

Figure 6.38 Climatic Data – Monthly Temperature, Relative Humidity

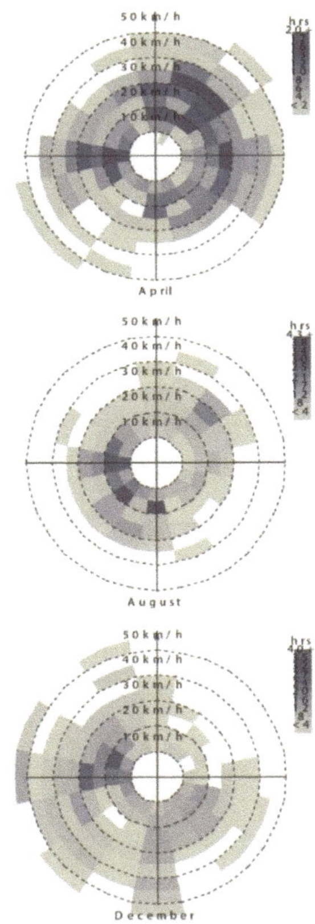

Figure 6.39 Prevailing Winds

curtain wall on the west side, enclosing a 3 ft wide cavity, and a single-skin curtain wall to the east, facing the lake. Three floors of open and partitioned study areas form "shelves," which are flanked by limestone precast "bookends" that house classrooms. The ground floor includes a café and connecting links to the existing library and chapel.

DESIGN INTENT AND BUILDING FEATURES

The new Information Commons (IC) is located in the historical heart of Loyola University Chicago's shore campus. By combining the latest computer technology with an open flexible design and innovative environmental systems, the new Information Commons is at the forefront of academic research facilities. The building not only is highly energy-efficient, but also provides a very encouraging space to study and research, while leading new ideas on campus in terms of sustainable architecture. The IC building house required spaces for new millennial undergraduates, including library, research and computer labs. The concept of the library has changed due to digitalization and globalization of information. Most libraries are equipped with high technology and resources. Loyola's IC building has about 222 computer workstations, wireless access, group study rooms, a café and classrooms, all of which have spectacular views of Lake Michigan and the surrounding campus. The building is located to the south of the existing Cudahy Library on valued campus green space with uninterrupted views of Lake Michigan. The architect's idea was to design a building that would integrate a modern architectural vocabulary with the historic Art Deco buildings on each side, while preserving the views of the lake. Working with Loyola, SCB envisioned the concept of a transparent, open computer space using the clearest glass available and a visually delicate structure to maintain views from the campus through the building to the lake. To blend this modern campus addition with the Art Deco Cudahy Library to the north and the Madonna della Strada Chapel to the south, SCB

Figure 6.40 (a) Look into Double Skin Façade Cavity (b) Loyola Building

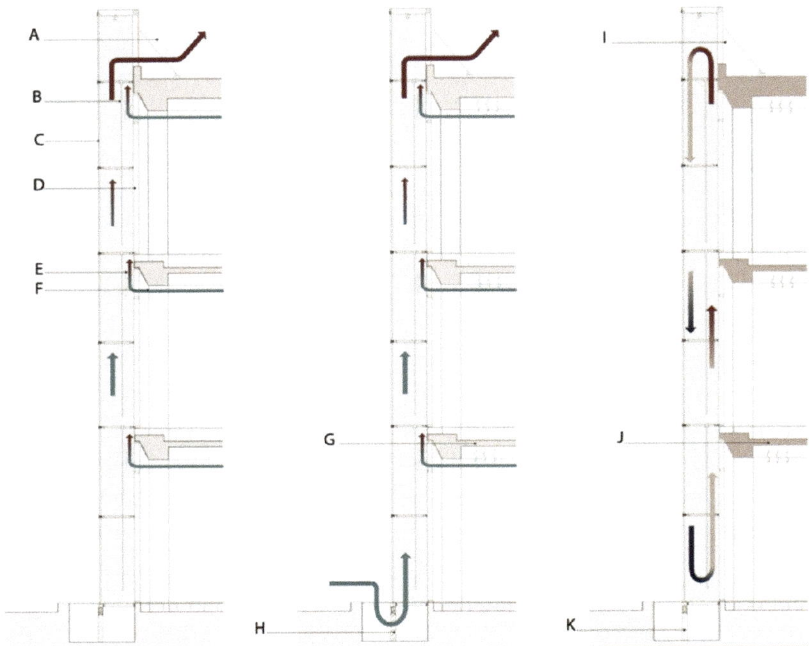

A	Operable awning windows
B	Automated blind system to control afternoon solar exposure
C	Outer skin/Clear Monolithic glass supported by a cable net structure
D	Inner skin/Mullion system
E	Motorized window open to the cavity
F	Natural air flow from opened east windows
G	Radiant cooling by cross linked polyethylene tubing set
H	Opened Louvers
I	Operable louvers closed in winter
J	Radiant heating
K	Closed louvers

Figure 6.41 Section Details

proposed containing the glass walls between limestone "bookends" that use materials, scale and details similar to those of the existing limestone buildings. The names of ancient philosophers and theologians are carved in the frieze atop the IC's bookends, further reinforcing the link with the historic buildings.

SCB proposed containing the glass walls between limestone "bookends" that use materials, scale, and details similar to those of the existing limestone buildings. The names of ancient philosophers and theologians are carved in the frieze atop the IC's bookends, further reinforcing the link with the historic buildings.

The exterior skin is an innovative application of a two-way cable-net support structure. The crossing network of 29 tensioned vertical and six horizontal stainless-steel cables support the 5-ft wide by 8-ft tall monolithic glass units at each corner with stainless steel clamping components. The slender cables along with a minimal field-applied silicone seal maximize the diaphanous appearance of the skin.

DESIGN GOAL

The goal was to build an environmentally conscious building that was complementary to Loyola's campus surroundings," said SCB chairman John Lahey. "We designed the Klarchek Information Commons using some of the most cutting-edge, sustainable technologies, which helped us to achieve LEED Silver certification for the building." The building's form, space, function, materials and construction all act to enforce and support each other in creating an environment that maximizes comfort while reducing energy consumption by approximately 50%.

One major goal of this study is to do a Building Performance Verification to compare the simulation result with the actual building energy consumption data provided in the *High-Performance Building Magazine*, Spring 2010; and in doing that, also checking the simulation tool capabilities. The second, is to change the building operation schedule in the simulation tool while keeping all the other input parameters as they are and check the effect of the building operation schedule on the traditional techniques for control of microclimates within this type of environment, including preferential glazing to admit or block insulation, appropriate location and orientation of spaces to introduce air currents within inhabited spaces, and employment of passive strategies (ducts, wind towers and shafts) to promote circulation, as well as heat extraction.

For Chicago's climate, the main advantage is improved thermal insulation. During the winter months, exterior skin increases external heat transfer resistance, therefore utilizing interior air for preheating the air cavity is advantageous. During the summer, air must be extracted in order not to cause overheating, by natural, hybrid, or mechanical modes.

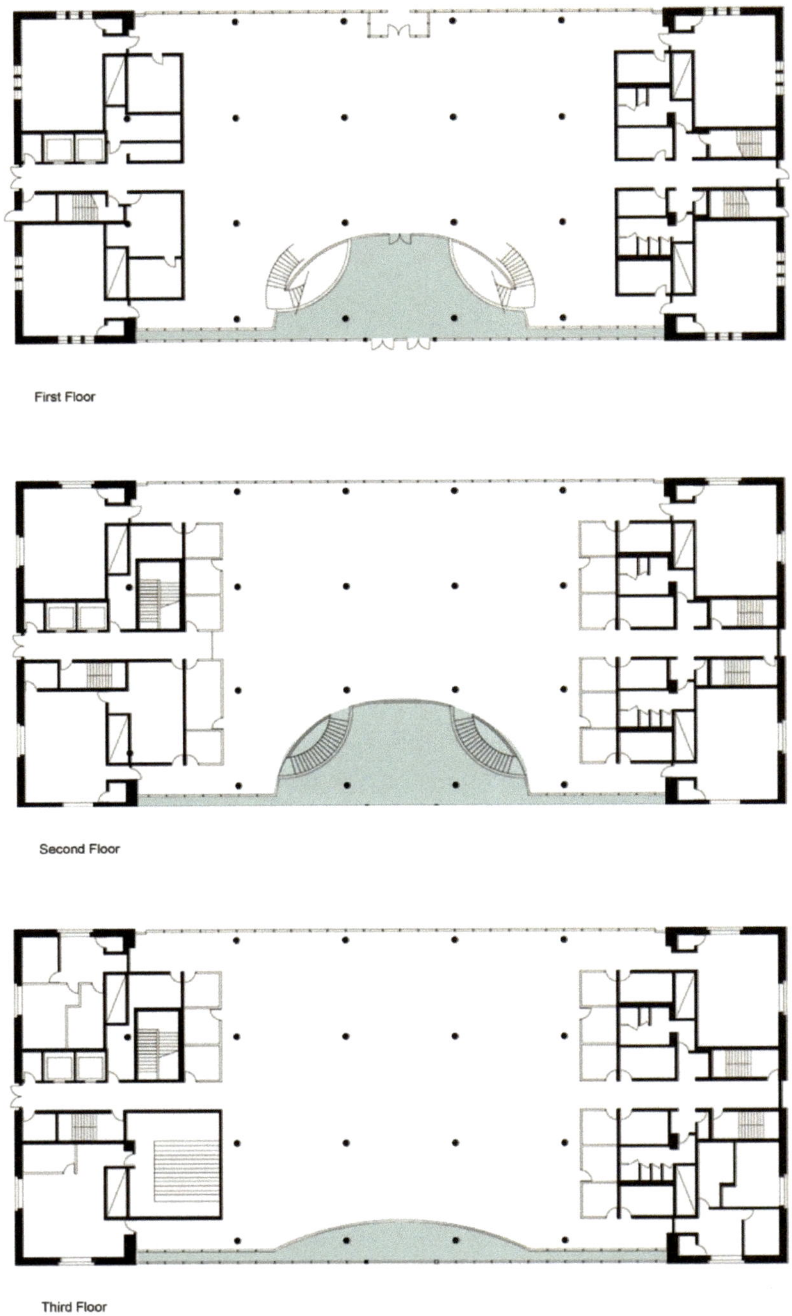

Figure 6.42 Plan Drawings (from the top), First, Second, and Third Floor Plans

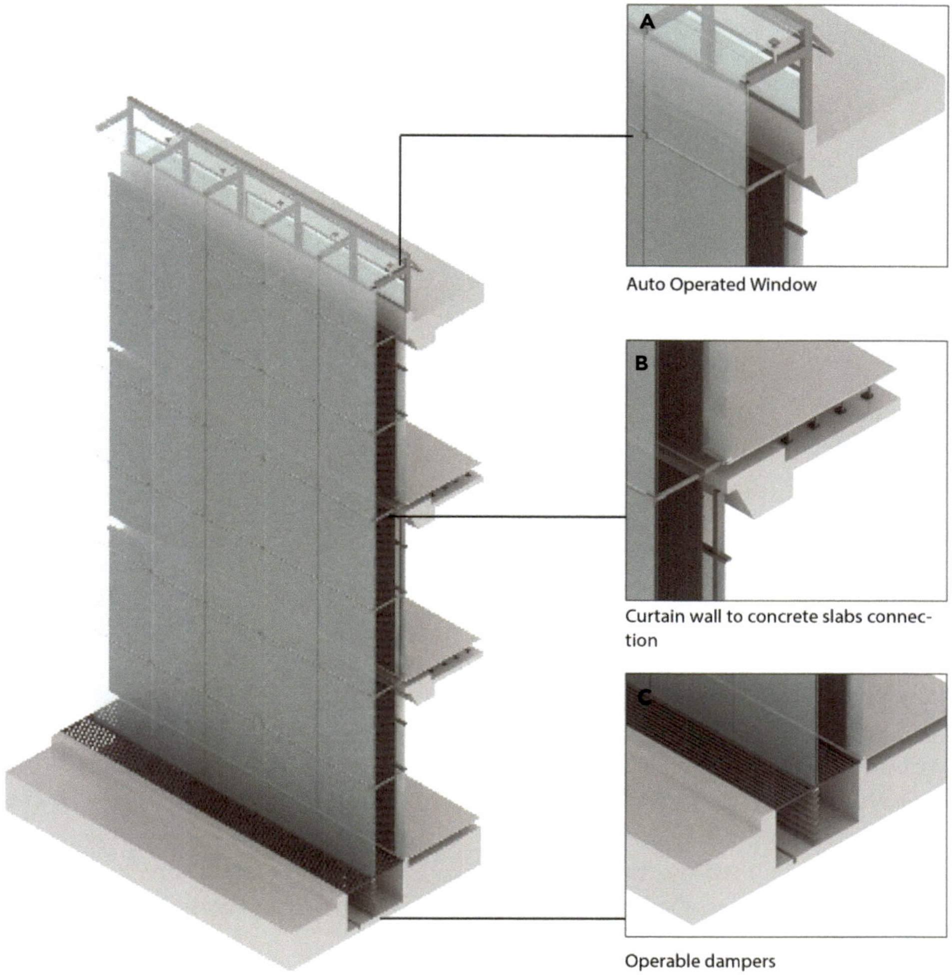

Figure 6.43 3D Section (a) Auto-Operated Dampers (b) Connection to Slab (c) Ground Dampers

KEY SUSTAINABLE FEATURES

- Chicago's first double skin façade to provide passive management of heat flow and ventilation, interstitial shading, and energy recovery
- Within this façade is a mechanically operable blind that adjusts daylight levels and heat transmittance from the west sun
- Daylighting: active, automatically controlled daylight harvesting. Windows, dampers, shades and blinds all respond to this climate data and adjust automatically
- Individual controls: individual thermostats for classrooms and study rooms in "bookends," as well as occupant operable windows in "bookends;" study rooms in open areas controlled independently

Figure 6.44 Roof Dampers

- Building systems are controlled by an array of sensors monitoring the temperature, humidity, and CO2 levels within the building, while additional sensors factor the outside conditions
- The climate control system that uses heated and chilled water delivered from existing campus facilities
- Water conservation: ultra low-flow urinals (one-eighth gallon per flush), dual-flush toilets and low-flow lavatories
- Radiant concrete slab ceilings provide thermal mass to cool in the summer and heat in the winter
- A green roof that collects rainwater and mitigates run-off to Lake Michigan

BUILDING ENVELOPE

ROOF: Type Concrete/insulated and vegetative

Overall R-value 40

Reflectivity 0.4

WALLS: Type Insulated precast concrete panels with additional insulation Overall R-value 13.7

Glazing percentage 48 Basement/ Foundation

Slab edge insulation R-value 10 Under slab insulation R-value 10

WINDOWS

1. U-value

Double façade 0.23

Other 0.46

2. Solar Heat Gain Coefficient (SHGC)

Double façade 0.49

East open area 0.35

Other 0.27

3. Visual Transmittance

Double façade 0.64

East open area 0.63

Other 0.45

LOCATION: Latitude 42, Longitude -87.656

Orientation West (double) façade 252° off north

BUILDING PERFORMANCE ANALYSIS

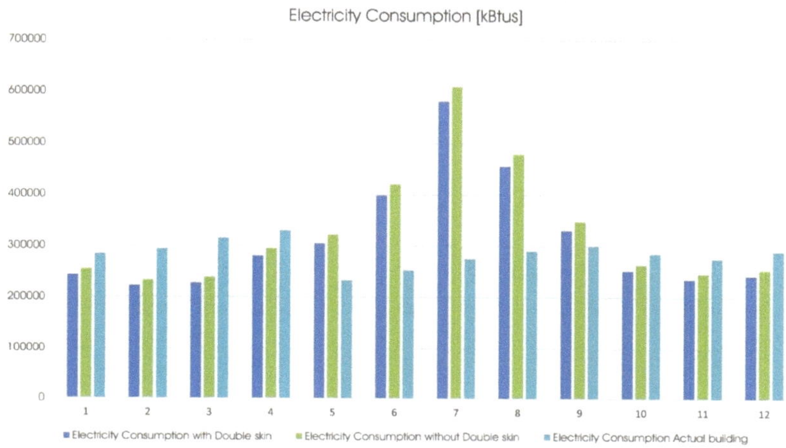

Figure 6.45 Electricity Consumption

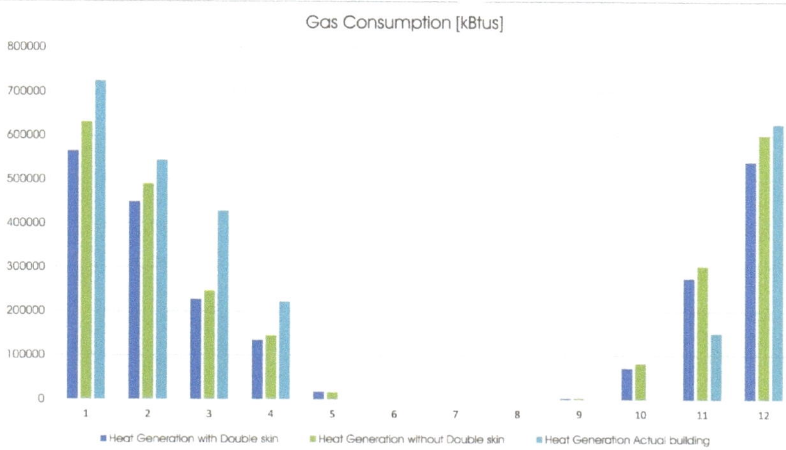

Figure 6.46 Gas Consumption

CONCLUSION

Simulation Output data according to all the input data taken from the building's available information, is compared with the actual building energy consumption result. This performance verification result is only off by 4.5%. This proves the Simulation tool capabilities of DesignBuilder in predicting

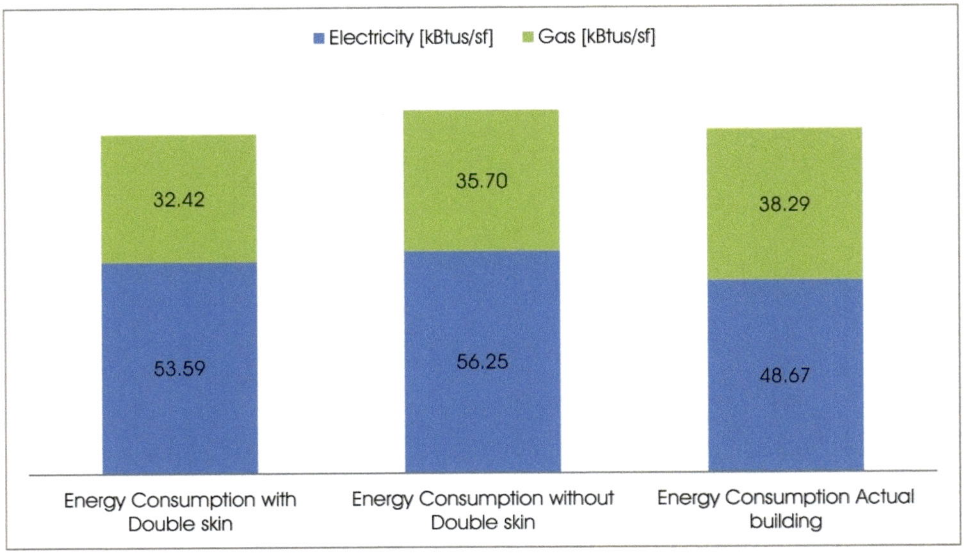

Figure 6.47 Energy Consumption Comparisons

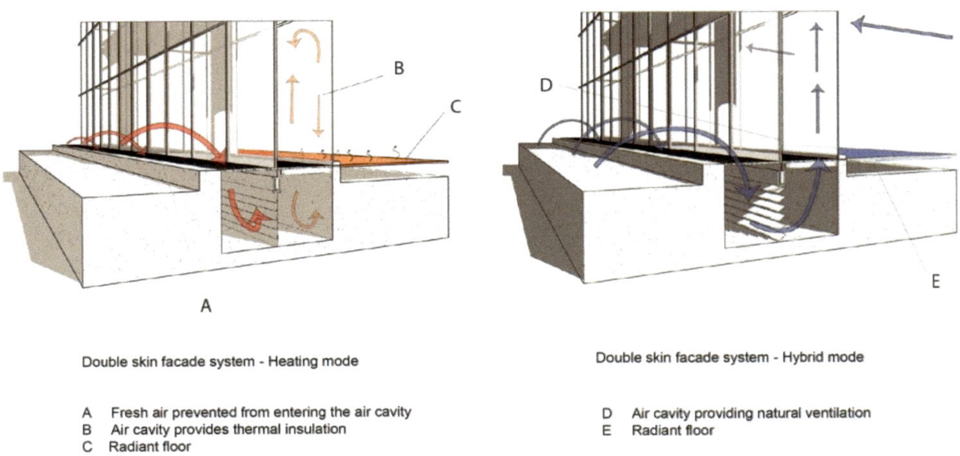

Figure 6.48 Detail Sections: Left: Heating Mode, Right: Hybrid Mode

the building consumption after the building is occupied and running. This study also concludes that for Chicago climate types there is an increase in the performance of the double skin façade compared to single skin. Therefore, if this tool is used beforehand at the design development stage, then one can predict the building consumption and with some changes and trials to make the building perform better.

In addition, design strategies for double skin façades should reflect the climatic conditions. For Chicago, reduction of heating loads is the primary concern, as well as balance with summer cooling load. Based on the results of this study, there are advantages to changing the air flow mode according to different seasons. In winter, there are advantages in utilizing trapped air to improve insulation and heat transfer between the exterior and interior. In summer, ventilation of the air cavity is essential for reducing cooling loads. The location of double glazing on the exterior skin improves the overall energy consumption.

INTERVIEW

Did you look at the single skin curtain wall system to be integrated into your building?

Yes, we did investigate using a single skin curtain wall system. However, we ultimately decided to use a double skin façade because the building is oriented north. Through analysis of the building and site solar gain, the amount of heat on both sides of the building was determined and the west side of the building was found to have a much higher solar gain than the east. We initially developed a shading system that is automated on the exterior with a single façade to protect against solar gain. One of our main goals was to incorporate a transparent façade system with views of Lake Michigan in one direction and the university quad in the other direction. It is very important to the university that we maintain lake views from the building because the building's current location previously served as a green space where students could sit near the lake and enjoy its beauty. As a result of this request, we intentionally maintained a visual connection to the lake through the building and designed the building to have a similar atmosphere to the lawn year-round.

Did you look at other types of DSF?

The double skin façade in use is partially active. The building operates in four modes. We designed the structure to take full advantage of natural ventilation. As we began to look at the natural assets of the site, we realized that there is a certain percentage of the year, roughly 20–30 %, that the windows in the building can be opened and mechanical systems can be turned off. Additionally, the site condition dictated that the building needed to connect to an existing library and chapel. Therefore, the building acts as a bridge between these two projects. With regards to the lake, it is angled at a degree atypical of the Chicago north-south grid. Due to this positioning, the location and transparency of our project were predetermined, and our main design goal was to create an efficient energy performance system for the building skin.

We examined the possible use of a double skin façade on both sides of the structure and studied the ways in which external shading might be useful in Chicago's climate. However, the University was concerned about the maintenance of these systems and the associated long-term viability, given the possibility of snow and ice on the exterior skin. With the University's main concern being transparency, we concluded that an operable shade would be the most efficient system to both block unwanted light, but allow views off to the exterior. In this way, the building's shading and transparency is flexible and can change throughout the day.

How does it perform?

We spent a large amount of money implementing the double façade and worked to utilize it as a tool to ventilate the building. The building operates in three modes: natural cooling, natural heating and a hybrid mode which incorporates both natural and mechanical cooling. Currently, our main concern relates to humidity. The levels of humidity in the structure are fully dependent on the operable windows and the entire system is fully automated. A weather station on the roof records data that is then received from a computer which controls how to position shades, windows, and mechanical systems. Additionally, there is a dual mechanical system present and a radiant ceiling which provides 90% of the heating through the ceiling. The air under the floor is used for ventilation and control of the temperature in the interior. The floor's purpose in regards to air is primarily displacement.

While there is very little displacement, heating, and cooling in this building, the primary heating and cooling surfaces take place in the concrete, vaulted ceilings. The concrete ceiling contains specs to weaken the shape of the vault for heating.

Radiant systems are typically more efficient when visible. Similar to how the sun provides radiant heat, the ceilings of interior spaces are responsible for the radiated systems. This system is opposite to convection technologies, where heat rises. In contrast to convection systems, radiant systems use electromagnetic waves aimed at the floor surfaces. Therefore, the floors are heated by a system above them. An example of this heat exchange can be seen in the library space. Any inhabitants sitting at a table in the library will naturally begin to generate heat in the chair they are sitting in. This heat will transfer to the underside of the table and eventually create a heat exchange between individuals in the space.

During the heating mode, heat is pushed towards the interior while during the cooling mode, heat is pulled from the interior. This system creates an operative temperature inside the building that combines air and radiant to create a desired level of comfort. We chose to use concrete in the building because it has the ability to store heat for cooling and act as a thermal mass when the building is in natural ventilation mode. With the use of concrete, even on a hot day, the ceiling remains cool. This allows for higher temperatures from outside to be brought in while the system is working in natural mode. While using natural ventilation, the building can reach 74 degrees.

When the double façade closes, ventilated air enters from the floor while the heating enters from the ceiling. The sun heats the perimeter area of the windows, creating a warm buffer between the outside air and the occupied space inside. For example, if it is 0 degrees Fahrenheit outside, the buffer area will most likely reach about 30 to 40 degrees Fahrenheit while the interior is kept at a constant unaffected temperature. With regards to the size and function of the double skin façade, the space between the two faces amounts to 3 feet. Due to the natural buoyancy of the air that is pushed through the façades, 3 feet is the minimum allowed distance that was allowed by our climate engineer.

These concepts are much more acceptable in Europe, as we often base our ventilation decisions on other factors in the US. Economic factors and code compliance are both motivators in the US that pushes many designers to remain with traditional systems. However, this building is 50% more efficient than the code requirement. The code requirement represents the minimum standards, and we aimed to achieve a much higher standard of efficiency in our structure than the code called for.

Did the consultant also model the double schemes of different modes of operation?

During the winter months, the windows remain closed and act as a thermal buffer, with the interior spaces conditioned from the inside. There is only a single skin façade on the east side of the building. As we began our analysis of how the sun affects this side of the building, we realized it did not require a double façade. At around 10 o'clock in the summertime, the sun is no longer shining on the east façade. Therefore, it is not necessary to place a sophisticated double façade on the east side of the building. This side of the building contains an insulated layered glass, fabricated of high-performance insulated glass. It also contains a shade between the two pieces of glass.

The shade contains a low-E coating, atypical of a standard shading device. The low-E coating is pointed towards the inside of the building, which traps heat between the shade and the façade. An operable window is located at the top of the shades and expels stale air in the morning. When the sun rises, the shades lower and prevent glare from entering the building. Functioning synchronously, the double façade brings air into the building and expels it naturally. There are no mechanical means in use. The building is designed to protect the interior from unnecessary solar heat.

Often, when designers begin to examine double skin façades, they do understand the physics and the science behind the system they are proposing. In comparison, our team had completed the necessary research and determined that the single façade on the east would create a positive flow of air from one side of the building to the other side, creating a natural ventilation mode for the structure. Dampers are located at the base and top of the façade, allowing for natural ventilation. Air is pulled through the façade due to natural buoyancy and is assisted by air flow over the top of the system. The façade itself extends eight feet above the roof, pulling out air through the building that was taken in from the east side. We studied prevailing wind patterns and created a diagram that displays the direction of airflow from the south and west. As air flows over the top of the pyramid structure, it creates a negative subject on the backside of the façade which helps induce the natural ventilation discussed previously.

With the windows open, cross ventilation occurs. The temperature of the interior spaces is able to be controlled by half a degree through this ventilation and façade system. The windows are capable of being opened and closed at different intervals. The computer has the ability to react to the natural environment, and is tuned to take advantage of that situation when all mechanical systems are off.

During the summer months, the dampers on the bottom of the façade are utilized. The metal shades located inside the double façade consist of metal horizontal blinds, each four inches wide. These blinds extend to protect the building from the sun. The highest heat gain on the exterior façade occurs in March. During summer months like March, these metal blinds absorb heat from the sun and then reradiate this heat outward. With the damper open, exhaust air can be pulled out through the cavity. This prevents the cavity from overheating. The temperature in the cavity can vary from 50 degrees in the winter to 80 degrees in the summer. This cavity acts as a "swinging" space between the building and the environment.

In the summer months, when all windows are closed, ventilation, whether natural or mechanical, is required. When working in mechanical ventilation mode, the system has the ability to cool the ceiling to 68 degrees, creating a comfortable interior temperature. It took data from three seasons of heating, cooling and natural ventilation in the building for the University to realize that it was going to take one year to work out the initial "bugs" in the building. The main issue with the façade during the first year of operation was the actuators and operable design of the windows.

The building performed very closely to the modeling and production that was initially predicted. For this, we are very proud of the team and their collaboration. There have been no issues with the performance of the façade system. The only aspect of the building we would have done differently is taking the measurement of verification. We initially perceived the credit to be too expensive and found it difficult to monitor the performance of the building, but looking back, I would now like to have this credit verified.

During the first year, as we worked to fix all the "bugs" in the building, there was an unknown energy consumption we could not pinpoint. It took time to evaluate where this energy was being used due to the way in which it was circulated.

In subsequent buildings, we are now placing measurement tools, mainly meters specifically for outlets, lights, mechanical equipment, and transformers. With all these elements metered separately, instead of linked to a single meter, it is easier to tune the building more efficiently. This was an important lesson we learned and will keep in mind in future projects. Around the same time that we first realized there was an unknown energy consumption, the IT department bought new computers for the building. However, energy consumption was not considered in these purchases. With our suggestion, the building is starting to use lower wattage computers. We are also working with the building on management aspects of the computers. They previously left all 300 computers running at all times, not leaving time for software updates. The computers are now put in sleep mode every night, consuming less energy and providing ample time for updates.

Overall, the success of this building is tremendous. Often, people enter the building and cannot pinpoint why they enjoy being inside the structure. It is truly a coalition of all the different aspects we have designed in the building that work to make a fascinating and enjoyable experience.

Ventilation

Create a positive flow of air from one side of the building to the other side of the building, and so a natural ventilation mode. There are physically a set of dampers at the base and a separate glass damper at the top here, and natural ventilation on those dampers, clothes in the bottom down their windows on the inner façade and the west or east façade open and airflow through the space naturally is pulled out through natural buoyancy and assisted with airflow over the top of this pair but it actually extends up eight feet above the roof there to create a draw of air to pull out the air through the building.

Cross Ventilation

Able to control the temperatures inside the space within half a degree using this system. Summer mode: dampers on the bottom are used because in the summer mode, the temperature outside is extremely hot and humid.

Sun Shading

Metal shades inside the double façade. Metal horizontal blinds are four inches wide and come down to protect the building when the sun is on that façade. for 52 days it is using natural ventilation. We had to simulate spaces in the building and how they would react to natural light and natural ventilation. We simulated natural light and ventilation and then used those to calibrate the two models so that they were able to incorporate those things in there.

HVAC

A lot of bugs to work out with the building's heating, cooling and natural ventilation system; it took a year to work these bugs out.

One issue faced was in the first year the actuators on the windows on the west and south side were having issues pushing the windows out as far as they were designed to. The inner façade is where we create the thermal barrier between the two.

Challenges

The tuning of the building was the main challenge. The building performed closely to the modeling; from a performance standpoint, everyone was happy.

It is not a good idea to take the "measurements for verification" credits because it was hard to monitor the operational building performance. In the first year, the building was consuming energy that no one could figure out where it was coming from. Started to put measurement tools into outlets, lights, mechanical equipment, etc, to better tune the building. Figured out that the IT department was leaving 300 plus computers on 24/7, which is where this excess energy consumption was happening.

Energy Savings

University: probably about a $2 million premium for the building, which we see through energy savings is about a 16-to-17-year payback for them. For an institutional owner that's going to own this building for 100 years, 17 years wasn't a big deal.

From a quality-of-life standpoint, the payback is huge. There were zero complaints about the comfort in the building. Now we're actually doing a survey of all the students to see where issues may be so that we're correcting those going forward because we've got five more projects, we're doing for the university that is incorporating a lot of the concepts of this building into their design.

The radiant provides most of the heating and cooling, 90%. We spent the money on the recovery system rather than trying to make it a double façade.

Sound isolation was an issue with a natural ventilation system. Occupants do not want to have noise coming in from a hall into a classroom because it will be destructive to people that are in the classroom. In this building, all the public spaces and those main collaboration spaces are naturally ventilated.

The classrooms are not naturally ventilated; the offices are naturally lit. They have windows, so they get their air through the façade.

The solar chimney activates natural ventilation for common spaces activated with a central atrium that creates the natural ventilation for a building. We gave ourselves a bigger zone for air temperature that had to be achieved for comfort and combined that with the radiant. This creates a dynamic building that takes advantage of natural ventilation in a much broader portion of the year than just on the few days when it's the perfect temperature outside.

Bibliography

Detail. Loyola University Chicago: Richard J. Klarchek Information Commons | *Enclos*. Accessed February 16, 2017. http://www.enclos.com/projects/loyola-university-chicago-richard-j-klarchek-information-commons.

McLauchlan, Donald J., and David Lavan. Case Study - The Richard J. Klarchek Information Commons. Efficiency by the Book. May 01, 2010. Accessed February 16, 2017. http://www.hpbmagazine.org/attachments/article/11988/10Sp-Richard- J-Klarchek-Information-Commons-Building-Chicago-IL.pdf

The Richard J. Klarchek Information Commons / Solomon Cordwell Buenz. *ArchDaily*. May 18, 2012. Accessed February 16, 2017. http://www.archdaily.com/236302/the-richard-j-klarchek-information-commons-solomon-cordwell-buenz

Figure 6.49 Biomedical Science Research Building

Biomedical Science Research Building

The University of Michigan

LOCATION
Ann Arbor, Michigan

BUILDING PROGRAM
Bio molecular research Labs, Offices

ARCHITECT Ennead Architects

SIZE
593,727 SF

COMPLETION DATE
2005

BACKGROUND AND CONTEXT

The five-story, 593,727 square foot, Biomedical laboratory of University of Michigan Ann Arbor, integrated sustainable design features with innovative mechanical systems to reduce the building's energy consumption. Although the initial construction cost for the double skin façade buildings is higher than conventional, energy efficiency measures allow for lower operating costs and a greater long-term rate of return. Nestled in the University of Michigan main campus in Ann Arbor, the Biomedical laboratory boasts 2 laboratory blocks, an office ribbon toward the south double skin façade, separated from the laboratory spaces by the atrium, a 300-seat auditorium, and a vivarium. Michigan has a continental climate, although there are two distinct regions. The southern and

Figure 6.50 Building on the Site

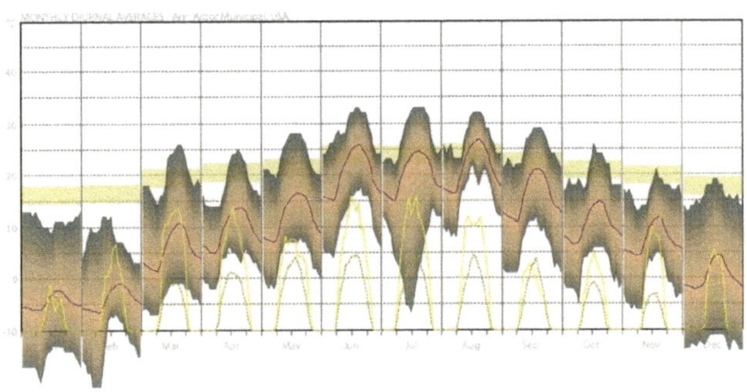

Figure 6.51 Hourly Temperature

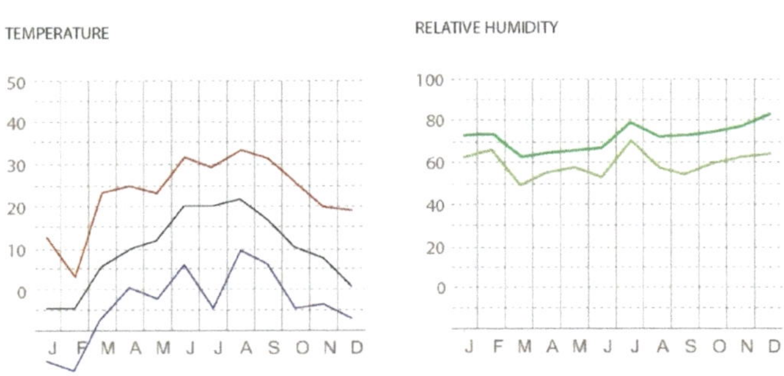

Figure 6.52 Climatic Data – Monthly Temperature, Relative Humidity

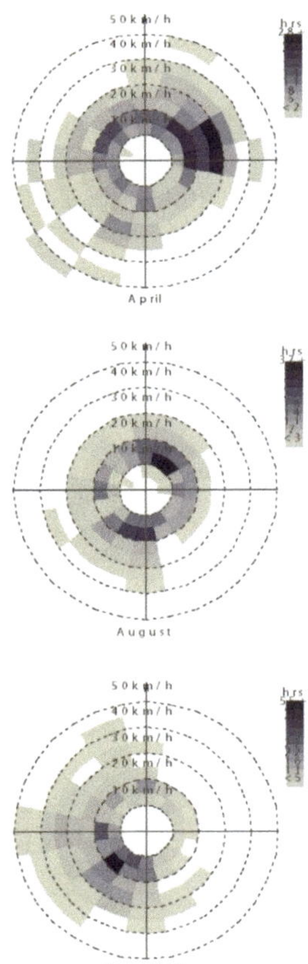

Figure 6.53 Prevailing Winds

central parts of the lower peninsula have a warmer climate, with hot summers and cold winters. The climate in Ann Arbor, Michigan is classified as a humid continental climate (Koppen Dfb with areas of Dfa). Summers are typically warm, rainy and humid, while winters are cold, windy and snowy. Spring and fall are usually mild, but conditions are widely varied, depending on wind direction and jet stream positioning.

The hottest month is July, with a mean temperature of 73.9 °F (23.3 °C). The coldest month is January, with a mean of 29.3 °F (1.5 °C). The city averages 42.5 inches (1,080 mm) of precipitation a year, with 41.8 inches (106 cm) of snowfall a year. Prevailing wind directions throughout the year are south-west and north-west.

DESIGN INTENT AND BUILDING FEATURES

Double glass façade for the full length of office ribbon: this passive façade includes a three-foot-wide space that provides a seasonal heating and cooling benefit to the building. During winter, dampers at the top of the double wall are closed. The heat from sunlight hitting the double wall is captured within this void, effectively reducing the heat load on the south face of the building. During summer, dampers at the top of the double wall are opened. The heat from sunlight hitting the double wall is flushed out via the stack effect, effectively reducing the cooling load on the south face of the building. And while the double glass façade is a cost-effective environmental design feature, it also produces a striking visual statement about the use of technology in a highly technical building.

- Exhaust air energy recovery;
- Use of specialized fan-coils in linear equipment corridors where equipment that generates a great deal of heat is located;
- Use of special sensors that automatically increase the volume of supply air when higher-than-desired levels of carbon dioxide are measured in returned air (non-lab areas only, since all lab air is exhausted); and
- Extensive use of sensors and remotely-monitored control systems.

Some of the building features are as follows:

A Unique High-Performance Façade:

- The double skin façade (two surfaces of glass, creating insulated air-space) is a multi-story (full height), full depth (3'), thermal flue. The façade allows for complete transparency while ensuring protection from excessive heat gain, heat loss, and glare.

Natural Light:

- The façade brings a significant amount of balanced natural light into the library, carefully controlled by the fixed and movable sunshades.

  Natural Ventilation:
- Operable windows in the façade allow for fresh air throughout the year (even in winter). In the winter, spring and fall, the windows allow heat from the cavity to be brought into the building. During the summer months the cavity is ventilated to the outside as an external air curtain.

Figure 6.54 View to Biomedical Science Research Building

The specifications of the project are:

A Unique High-Performance Façade:

- The double skin façade (two surfaces of glass, creating insulated airspace) is a multi-story (full height), full depth (3'), thermal flow. The façade allows for complete transparency while ensuring protection from excessive heat gain, heat loss, and glare:

BUILDING PERFORMANCE ANALYSIS

Energy Savings

- The cavity can be open in summer to keep heat from entering the building and closed in the winter to create an insulating "thermal blanket."

Natural Light

- The façade brings a significant amount of balanced natural light into the building, carefully controlled by fixed and movable sunshades.

Natural Ventilation

- The double skin façade ventilates to the outside during the shoulder seasons.
- Floor-to-ceiling vision glass allows natural daylight to flood the interior spaces and a stainless-steel screen forms a layer within the glass cavity.

The plan was divided into multiple zones according to geometry, proximity, and use. Each zone was then assigned an activity template (indicated by color in the images) which associated that zone

A Operable louvers / opened in summer
B Concrete floor on metal deck
C Spandrel panel with insulation
D Retractable blinds
E Inner adjustable blind
F Outer skin / Mullion system
G Low-E coated glass
H Vent / opened in summer

I Operable louvers / closed in winter
J Vent / closed in winter

Figure 6.55 Summer and Winter Façade Performance

Vent

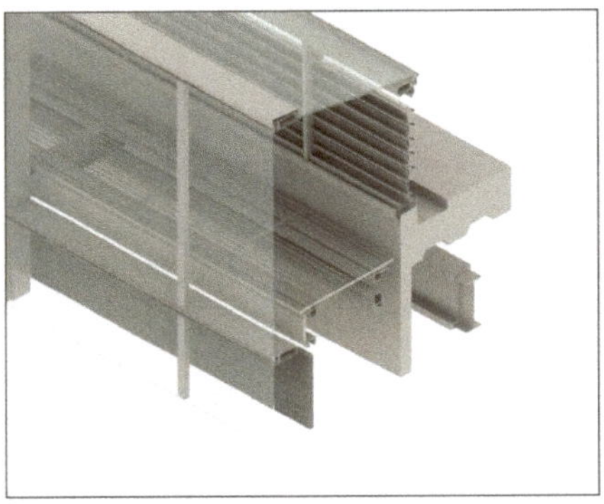
Operable Louvers

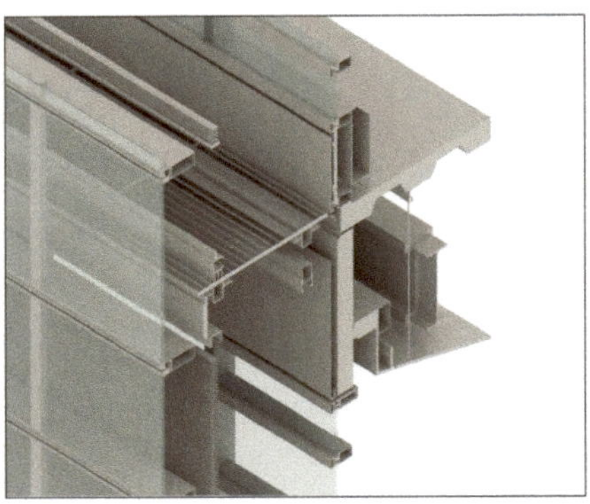
Retractable Blinds

Figure 6.56 3D Section and Details: Vent, Operable Louvers, and Blinds

with different energy profiles and schedules. As shown in Figure 6.57, the zoning and schedules are often repeated floor to floor. The atrium is treated as one single zone inside DesignBuilder, with openings in the floor and ceiling of the corresponding floors connecting them. The building has 3 floors and a basement, and the zones are divided by the use of each one. We have 8 different zones: hall/lecture/assembly area, circulation area, display and public area, generic office area, reception, toilet, workshop – small scale, and light plant room.

After obtaining basic building and weather data, a simulation of the building's energy performance was undertaken. This was the base case simulation for a single skin façade system. The simulation software used for this research was DesignBuilder with EnergyPlus as an engine.

Since the behavior of double skin façades is highly dependent on the type of climate, the following sections outline a process for selecting the type, analyzing characteristics and properties and selecting strategies preferable for a specific context. Components of the studied double skin included an internal layer composed of a 6 mm clear glass pane, a 20 mm argon-filled gap, and a 6 mm internal low-E glass and aluminum frame with thermal breaks. The external layer consisted of 6 mm clear glass. Components of the single skin were similar, with double glazing composed of 6 mm external clear glass, a 20 mm argon filled gap, internal 6 mm low-E glass and aluminum thermal break frame. Both cases included shading blinds, where in the case of double skin they were placed in the cavity, and for double glazing on the interior of the building.

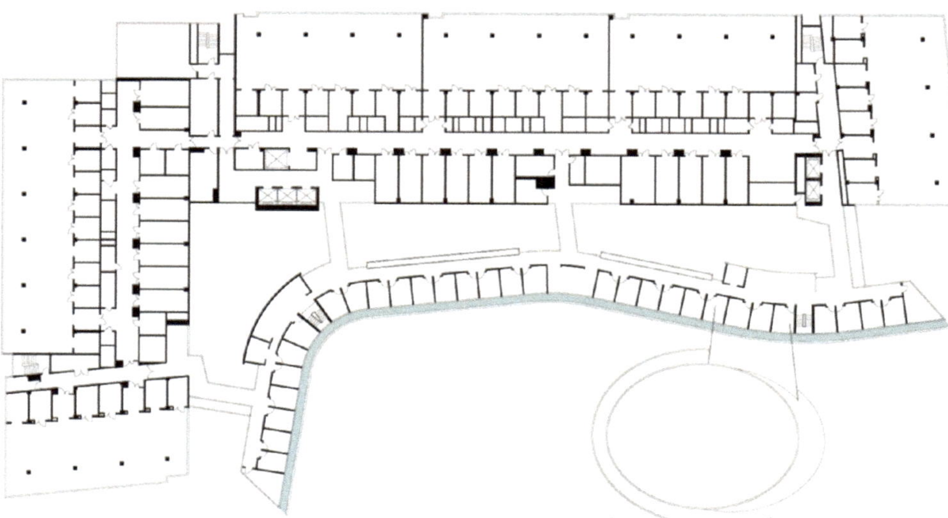

Figure 6.57 Typical Plan

Figure 6.58 View to Double Skin

Figure 6.59 Interior Atrium

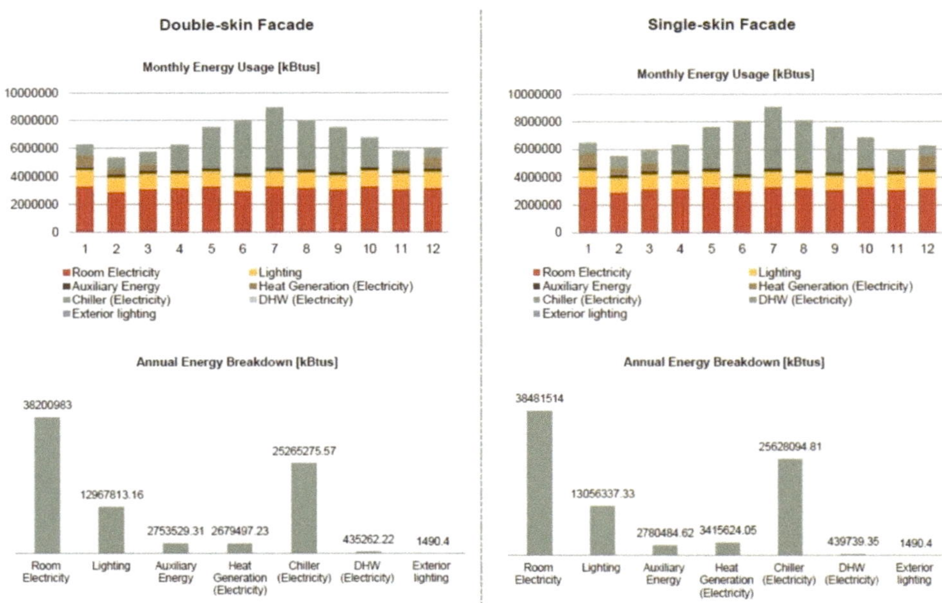

Figure 6.60 Energy Consumption Comparisons – Double and Single Skin Façade

The building will not only conserve significant amounts of energy, but it will also create a highly conducive environment for study and research while ushering in a new era of campus architecture focused on resource conservation.

The benefit of the double façade becomes apparent when comparing heating loads.

The annual energy consumption of the double skin has been compared with single skin as illustrated above. It was discovered that in total a double skin façade consumes 5 % less energy than a single skin façade.

One major goal of this study is to do a Building Performance Verification to compare the simulation result with the actual building energy consumption data. As illustrated in Figure 6.60, the modeled energy consumption and actual energy performance are within 10 percent range which is pretty accurate.

CONCLUSION

In general, energy consumption in buildings is determined by function, climate, building components, construction, control, and settings. The climate and the ambiance are considered as boundary conditions in energy simulation. Building function also has an important impact on energy use. Building components and construction both provide great potential for improvement of energy demand in such areas as adequate thermal insulation, a key component of energy consumption. A careful choice of windows and shading devices should help to avoid additional solar gains. Incorporating efficient HVAC equipment and heat recovery techniques may also reduce the energy use.

Designing a high-performance façade system will make a tremendous impact in minimizing energy consumption and optimizing the thermal condition. One major goal of this study is to do a Building Performance Verification to compare the simulation result with the actual building energy consumption data. The second major goal of the study is to change the building envelope properties in the simulation tool with keeping all the other input parameters as they are, to study the impact of the envelope on the control of microclimates withing this type of climate. These strategies include preferential glazing to admit or block insulation, appropriate location and orientation of spaces to introduce air currents within inhabited spaces, employment of passive strategies (ducts, wind towers and shafts) to promote circulation, as well as heat extraction.

Energy savings are not significant, but other aspects can justify the recourse to the double skin façade other than energy efficiency such as thermal comfort in all glass façade buildings which is not within the scope of this chapter. For this climate type, the main advantage is improved thermal insulation. During the winter months, exterior skin increases external heat transfer resistance, therefore utilizing interior air for preheating the air cavity is advantageous. During the summer, air must be extracted in order not to cause overheating, by natural, hybrid or mechanical modes.

INTERVIEW

This building required over 10 years to design and construct. Comparatively, the cost is very expensive when considering other buildings in the University. However, the University is willing to spend more money on this building because it acts as a signature building for the campus, standing out among the surroundings. Located on a primary road in Ann Arbor, those driving on the road receive exemplary views of the building heading the downtown area. Because the building is in plain view to many pedestrians, the exterior façade was very important. Since the building is primarily oriented north, the west façade receives dramatically higher levels of heat rays than the other sides. Therefore, a double façade was only necessary from this direction. Since its completion, the project has won a national AIA award. It is not simply our opinion that the building exemplifies smart design strategies, but the design community has concluded that the building has merit as well.

When discussing designs, the Dean of the College for the Medical School told us that he wanted a building that is different from other buildings that the medical school has constructed. His primary wish was for his faculty to be able to easily consolidate and get together. This is a common request from medical researchers who often tell us they would like to be surrounded by others in their lab environment. In the previous building, the researchers mainly worked in independent rooms. To create an entirely new and cohesive organization of the building, we paid close attention to the efficiency of the laboratory and office spaces and their connection to each other. The building design achieved the Dean's request through the intersection of technical and investigatory spaces.

The clients originally envisioned the shading features to be actively controlled with moveable sunscreens that are responsive to real time solar conditions. Although the performance of this type of system is very high, the practicality is far lower. Because the building is multiple stories, if implemented, this system would require thousands of small motors to control all the shades. The high price required to maintain each of these features, along with the high chance that multiple could frequently break, the design team decided to implement a more practical concept. Although the chosen concept is more practical and cost-effective, it does not save as much energy as the original proposal.

Did you look at other types of DSF, such as corridor type?

Yes. However, when we looked at the actual way that air travels in and out of these louvers, the University was not convinced that there would be adequate space for air flow from gravity without the additional use of a fan. The system we chose to implement utilizes a much larger open area and creates a greater stack effect compared to a corridor type system. In our system, the air is encouraged to move, not limited to do so. Based on our research, we believed that the larger the height of the façade, the larger the open area should be.

With regards to aesthetics, the architects preferred to not place large interrupting bands on each level. They requested that the façade either include smaller bands or none at all. We considered several options of double skin façades, but ultimately decided on this system of multi-story air transfer. Of the many systems we researched, one option included grilles at the top of the system. We decided to implement this, but to a much higher level. Another option that we studied recommended widening the open area from three to four feet. This we did not incorporate. A third option that we investigated involved how to make the system efficient at each individual floor. We did not incorporate this technique either. We were very drawn to the first option studied and based our system on its findings.

HVAC System: The heat recovery is only in the lab space and the office space does not need heat recovery. We can recirculate the air from the office.

The lab is at minimum 6 Air Change per Hour (ACH based on OSHA Standard. Office might be higher (20 ACH).

Heat recovery is air handling, which has water coils that take the heat. The water goes to supply an air fan to pre-temper the air and it may be boosted if needed. 100% of the air comes from the outside. For better, indoor air quality, the make-up system brings in the outside air to replace the air that is exhausted out.

Bibliography

Navvab, M. (2005). Full scale testing and computer simulation of a double skin façade building. 9th International IBPSA Conference Building Simulation 2005, Vol 2, Montreal (Canada). 831–838.

Safer, N., Woloszyn M., Roux J.J., & Kuznik F. (2005). Modeling of the double skin facades for building energy simulation: Radiative and convective heat transfers. 9th International IBPSA Conference Building Simulation 2005, Vol 2, Montreal (Canada). 1067–1074.

Stec, W., & Passen D. (2003). Defining the performance of the double skin façade with the use of the simulation model. 8th International IBPSA Conference Building Simulation 2003, Vol 2, Eindhoven (Netherlands). 1243–1250.

Figure 6.61 Seattle Justice Center

SEATTLE JUSTICE CENTER

LOCATION
Seattle, Washington

BUILDING PROGRAM
Court house and police headquarters

ARCHITECT
NBBJ

SIZE
297,678 Sf

COMPLETION DATE
2002

BACKGROUND AND CONTEXT

The design of the Seattle Justice Center encourages congregation and fulfills the long-standing need for a place of civic discourse for the city of Seattle. With the updated security systems and highly technological aspects of construction, the building stands for itself. This project joins different programs (the municipal courts and police station) in a shared location, achieving distinct individual needs. Each program has its own architectural identity in addition to its function. The result is a juxtaposition of mostly glass courts with a predominantly stone police station.

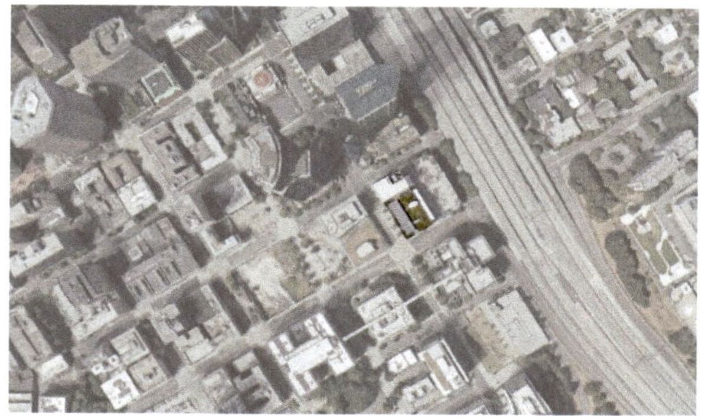

Figure 6.62 Building on the Site

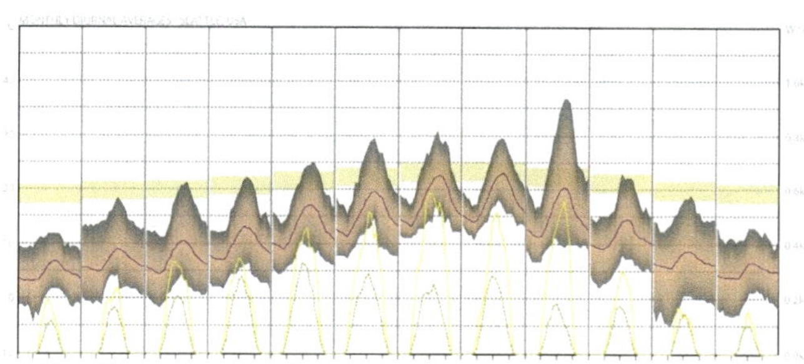

Figure 6.63 Hourly Temperature

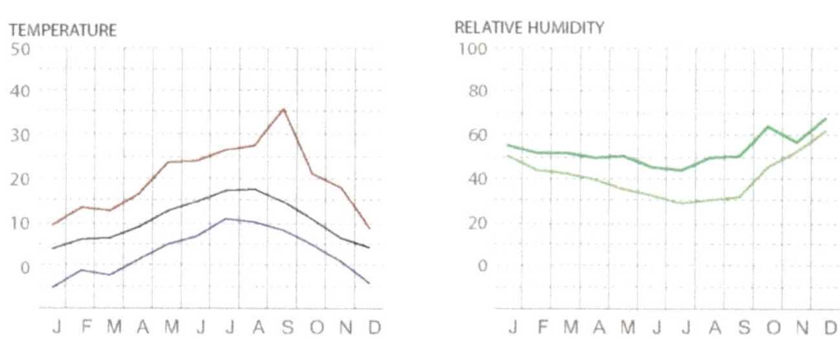

Figure 6.64 Climatic Data – Monthly Temperature, Relative Humidity

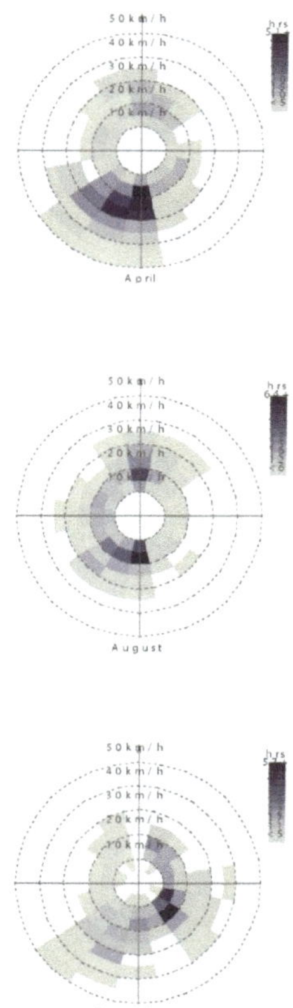

Figure 6.65 Prevailing Winds

DESIGN INTENT AND BUILDING FEATURES

The most notable sustainable features pursued in the design of this project are:

Glazed Thermal Buffer Wall

A large section of the west façade of the building was designed as a naturally vented "double skin" façade: two separate planes of glass separated by a 30-inch air space. The double curtain wall allows light penetration while minimizing heat gain. The air space has automatically

Figure 6.66 View to Double Skin

Figure 6.67 View to Cavity

Figure 6.68 View to Dampers on the Roof

Figure 6.69 View to DSF and Cavity

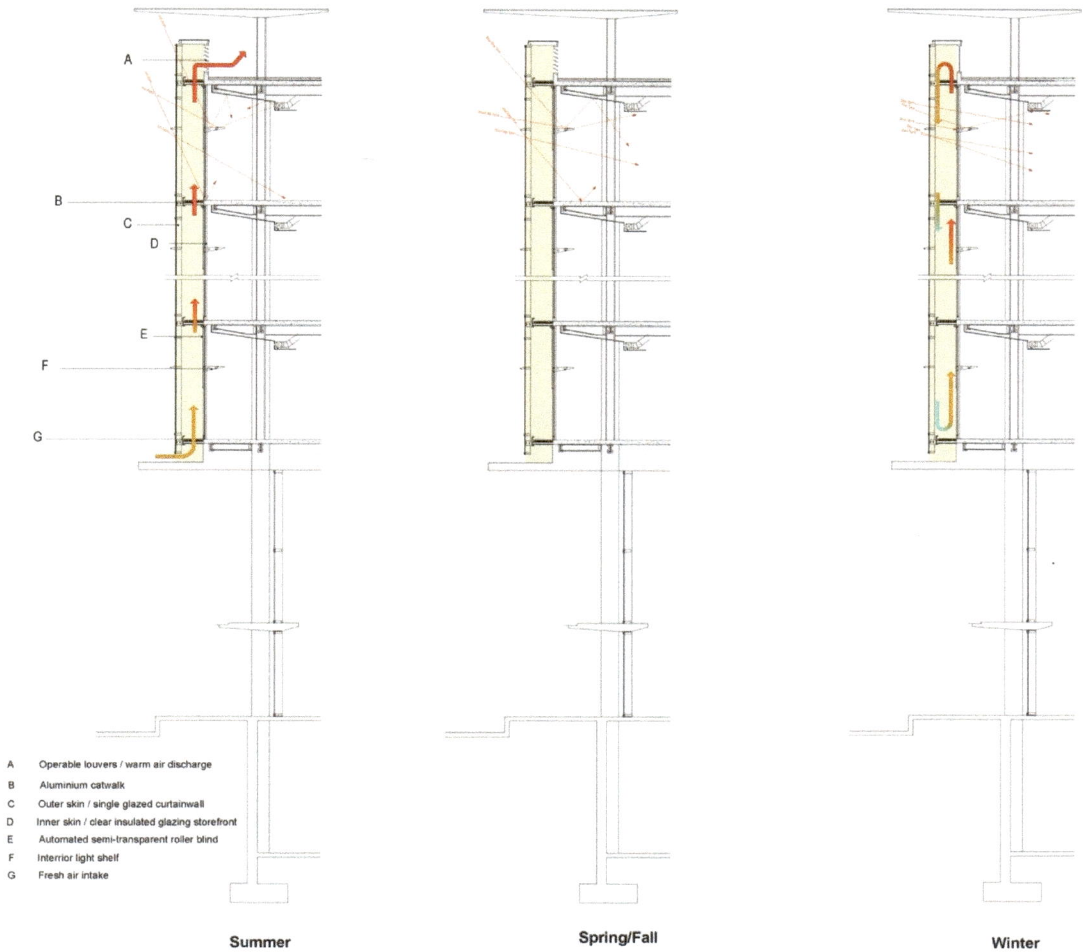

Figure 6.70 Section Performance of the Double Skin Façade in Different Seasons

Top Grille

Bottom louvered Grille

Figure 6.71 Section and Details: Top and Bottom Openings

controlled louvers at the roof level to release or retain heat as required. Light-shelves at each floor of the curtain wall reflect natural light to the interior as well as acting as shading devices.

Garden Roof

The principal roof area of the building is designed as a "green" roof and terrace with drought-resistant and low-maintenance plants. The planted roof adds an insulating layer of soil and removes solar heat gain through photosynthesis. It absorbs and stores rainwater and filters pollutants out of the air while returning some oxygen to the atmosphere.

Water Harvesting

Water harvesting is employed as part of a water detention system, storing rainwater for irrigation. Using harvested rainwater to water the public plaza plantings reduces the building's use of municipal potable water.

BUILDING PERFORMANCE ANALYSIS

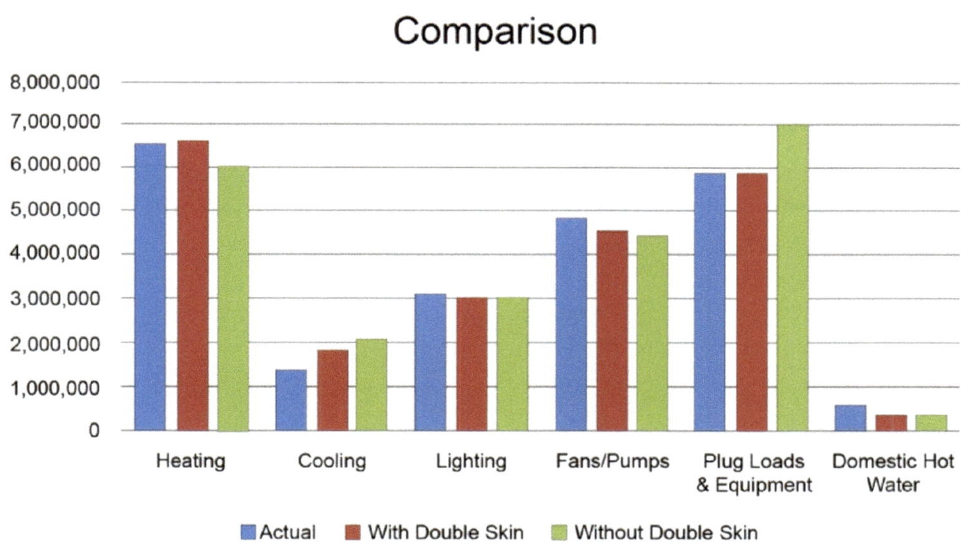

Figure 6.72 Energy Consumption Comparisons Double and Single Skin Façades

CONCLUSION

The Seattle Justice Center, which serves as a space for city court and police headquarters, was designed with high technology materials and sustainability to last for at least 100 years from date of construction.

The key features of the project are:

- The double skin façade, multi-story-type helps with the regulation of hot air into the building and maintaining the thermal comfort levels inside the building. This also keeps the prominent views from inside the building and glazing avoids viewing in from outside, providing high security.

- The green roof helps in the renewability of energy, and also keeps the building cooler.
- Light shelves provided at 8' from finished floor level at every floor, provide ambient lighting and also serve as a sunshade around midday.
- The 2' air-gap between the two panes of glass is obstructed with grille grating that patternizes the wind flow and ensures that the hot air exits through the vent provided at the terrace level.

INTERVIEW

Please give us a brief introduction to the project.

The Seattle Justice Center opened its doors in October 2002. The nine-story stack building utilizes louvers controlled by the building automation control system. When designing the structure, ARUP used CFD modeling to examine different configurations. In an attempt to create a seemingly simplistic design, we studied which type of glass to use and how wide the optimal gap should be. The cost was at the forefront of our design limitations, and thus we worked to use as many passive systems as possible. Because the building is located in Seattle, WA, we studied glare potential and solar heat loads at different times of the year. Our main concern was solar heat gain during the summer months.

How many parameters did you look at?

We looked at 30" clear between the two glass panels. Some parameters were dictated by OSHA with regards to access requirements.

To prevent heat from entering the building through the windows, the shading feature is mounted in the gap between the glass panels.

The core is meant to be as transparent and welcoming as possible, providing views of the outside.

The back half of the structure serves as a parking garage for the inhabitants of the building with the main façade directed southwest towards the water. In this way, the directional grid is slightly skewed towards the water. On the 12th floor of the structure, the double skin façade contains louvers which protect the façade, depending on the wind pattern. In contrast, the bottom of the façade is always open. When the environment is warm, the air rises and cannot escape if the louvers are closed. This is considered heating mode because the warm air is trapped inside the façade. When in cooling mode, the air is continually pushed through the façade.

The weather in Seattle is relatively temperate, so condensation is not a concern. With regards to solar heat gain, the catwalk acts as light shelves to protect individuals who are positioned close to the exterior glass.

Benson is based in Portland, OR. They fabricate custom curtain wall systems and often pair with Kawneer. We hired Benson to create the custom exterior layer of our system. This consists of a tube steel frame around glass panels. Each panel of glass is 9' wide and 7' high. In contrast, the interior system consists of standard shelf system panels provided by

Kawneer. The insulating glass is a monolithic panel which is coated in low-E to keep the cost down. While remaining cost-effective, these insulated monolithic glass panels have a relatively high level of solar efficiency.

Can you elaborate on the cost aspect of the system?

With regards to the overall cost, we conducted a dim control and solar control study to determine the cost-efficiency. We determined that a temperature sensor on floors 3, 6 and 9 would be more effective for controlling the louvers.

The principal roof areas of the building are designed with a "green" roof implemented. These "green" roofs contain drought-resistant and low-maintenance plants. This concept inherently requires an insulating layer of soil and removes solar heat gain through photosynthesis. It absorbs and stores rainwater, which is absorbed from the "green" roof and then stored before being filtered. When pollutants are filtered from the water, some oxygen is additionally filtered and returned to the atmosphere. The harvested rainwater is used to irrigate plants as needed and to charge the water feature in the landscape design. Using harvested rainwater reduces the building's use of municipal potable water.

Bibliography

Case Study by the Cascadia Region Green Building Council, Seattle Justice Center (n.d.). City of Seattle Department of Design, Construction & Land Use, "Case Study: Sustainable Building, City of Seattle Justice Center," dcluINFO (February 2003)

Gonchar, J. (2003). Glass facades go beyond skin deep. *Engineering News Record*, February 10.

Llona, J. (2004). Justice Center Takes the LEED. *Seattle Daily Journal of Commerce*, April 23.

NBBJ. (n.d.) Naturally Vented, Double-Skin Façade – Thermal Flue Façade, documentation on the Seattle Justice Center.

Figure 6.73 Eli and Edythe Broad CIRM Center for Regenerative Medicine at USC

Eli and Edythe Broad CIRM Center for Regenerative Medicine and Stem Cell Research at USC

LOCATION
Los Angeles, California

BUILDING PROGRAM
University Research Building

ARCHITECT
AGF Architects LLP.

SIZE
91,485 SF - Five Stories

COMPLETION DATE
September 2010

BACKGROUND AND CONTEXT

The Eli and Edythe Broad CIRM Center for Regenerative Medicine and Stem Cell Research was designed as a way to reinforce the existing relationship between the research and clinical programs within the University's Health Sciences campus. The building was oriented and designed with narrow floor plates in order to respond to the sun, which also allows for the maximization of natural daylighting.

The site is very constricted, located on a sloping site that is bounded by two streets, a courtyard, and the adjacent

Figure 6.74 Building on the Site

Zilkha Neurogenetic Institute (ZNI). The design of the site includes innovative solutions for water runoff and drought

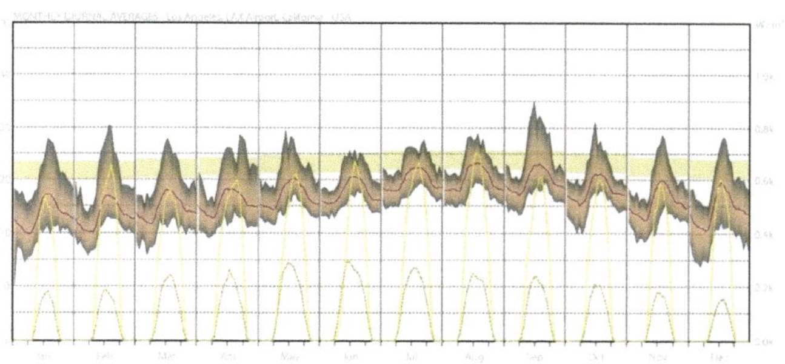

Figure 6.75 Hourly Temperature

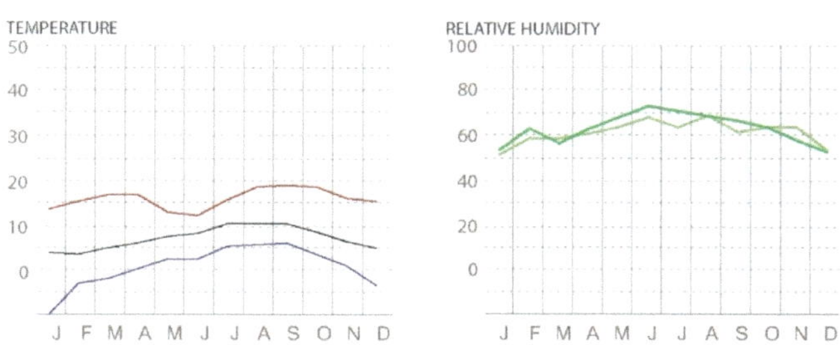

Figure 6.76 Climatic Data – Monthly Temperature, Relative Humidity

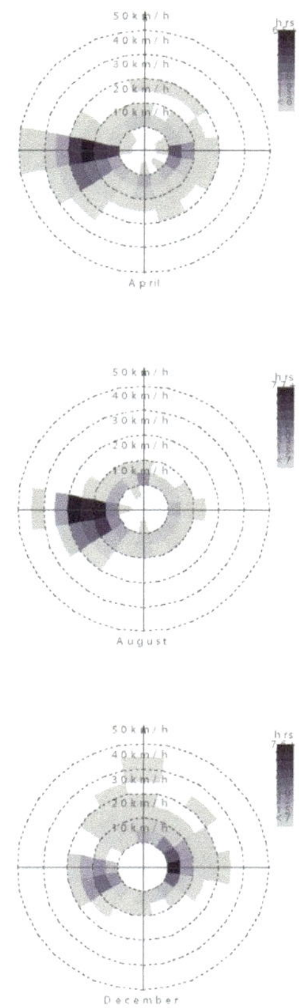

Figure 6.77 Prevailing Winds

tolerant landscape design. Water collected from the roof is filtered through planters before it leaves the site.

DESIGN INTENT AND BUILDING FEATURES

The design seeks to capture the boundlessness of science throughout the building. The idea was to support and express the dynamic nature of science which thrives on open inquiry and change. The minimalist quality of the design and materials is expressed by the firmness of black granite, stainless steel and transparent glass, all of which rest on a flagstone platform. The rectangular form shows openness through the use of glass façades spanning the long sides of the building. These double as an effort to maximize natural daylighting in the offices and laboratories, allowing the light to flow freely through the building. The ability to do this is accomplished by innovative glazing systems that respond to the sun's orientation by controlling heat gain and minimizing glare. The east façade also features a ventilated, double glass wall which acts as a buffer to moderate the temperatures of the interior spaces."

Figure 6.78 Exterior Façade

Figure 6.79 Exterior Façade

Figure 6.80 Bottom Grille

Figure 6.81 Double Skin Façade Detail

The design includes an interior skin with low-E coated insulated glazing supported by a unitized aluminum curtain wall system. The exterior laminated glass is cable-suspended from the overhead, cantilevered building structure.

The vertical cables are supported by T-shaped outriggers at each floor level to minimize cable span and deflections under loading. These steel outriggers double as support for maintenance grating and shading devices located within the cavity. Point-fixed bolted laminated glass attaches on a 5'3" vertical grid to the cables. The two skins enclose a 3-ft deep by 68-ft tall cavity, which acts as an outdoor air curtain by bringing fresh air into the lower cavity and using the stack effect to exhaust warm air through a series of automated, louvered vents at the parapet.

These operable louvers allow the cavity to be opened or closed to optimize cavity temperature and airflow, which will reduce undesirable conductive gains and losses through the interior glass. The louvers are controlled by comfort temperature sensors located on the interior surface of the interior façade. Though considered in schematic design, natural ventilation is not introduced to the interior spaces due to the sensitivity of the laboratory environment.

The CIRM building's multi-story double skin façade is located on the southeast envelope of the new stem research

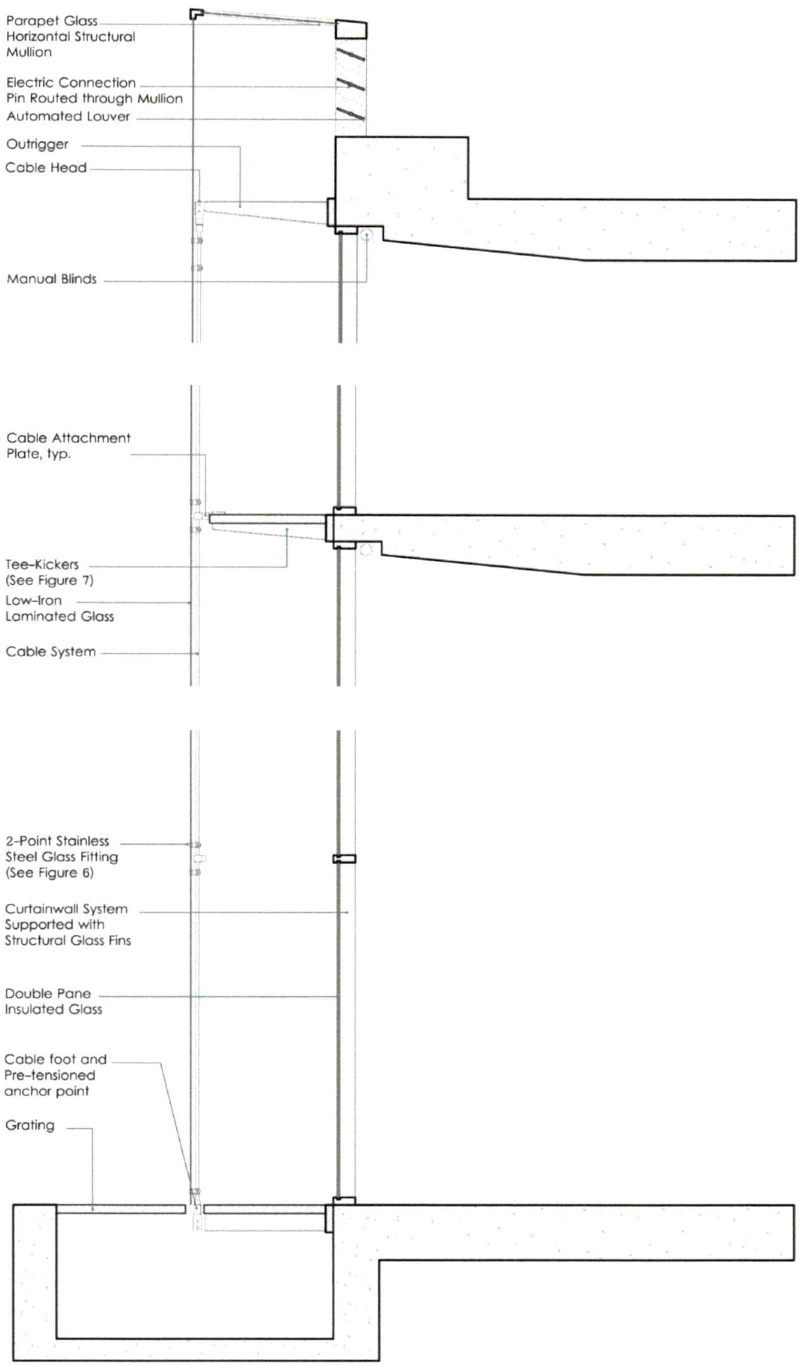

Figure 6.82 Detail Section of DSF

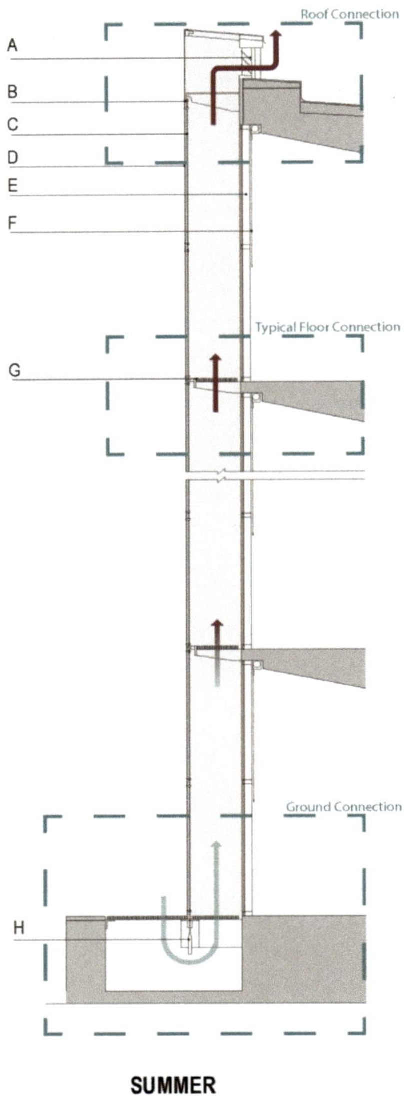

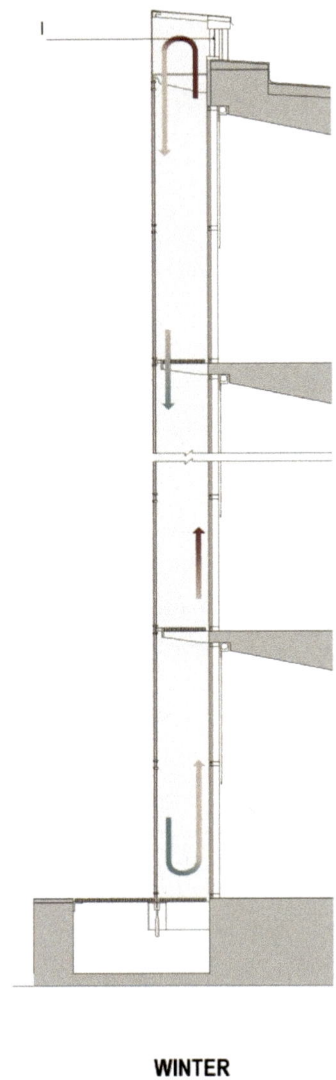

A Automated louvers B Cable Head
C Cable Systems
D Outer skin / Low-iron laminated glass
E Inner skin / Curtain wall system with Insulated glass
F Manual Blinds
G Grille
H Cable and pre-tensioned anchor

I Closed louvers in winter

Figure 6.83 Section of Double Skin Façade and Its Performance in Summer and Winter

A vertical cable mullion system defines the exterior double skin façade. Steel cables that are attached to kickers, which tie back to cantilevered steel outriggers, provide lateral support to the façade and support the glass.

The cable is attached to the end of a kicker extending From the floor slab.

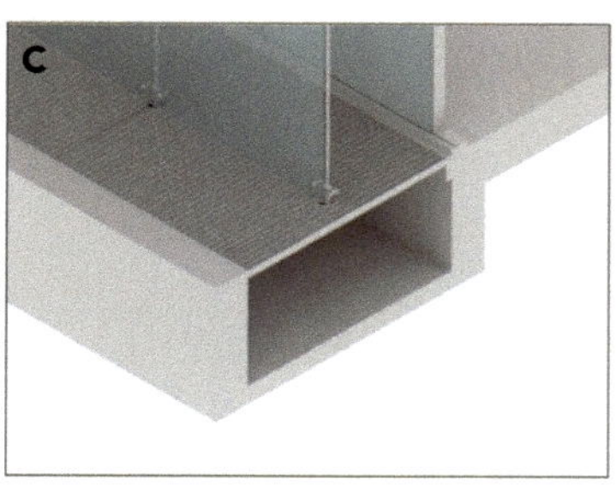

Outboard glass connection detail

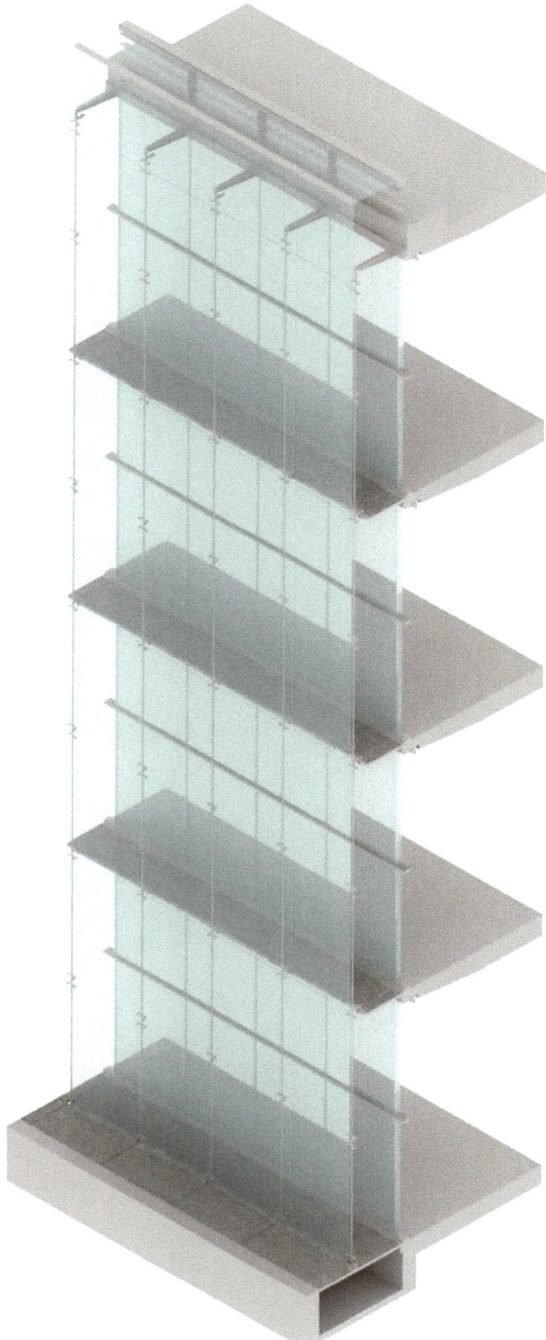

Figure 6.84 3D Section and Details (a) Top Grille (b) Bottom Grille (c) DSF Connection

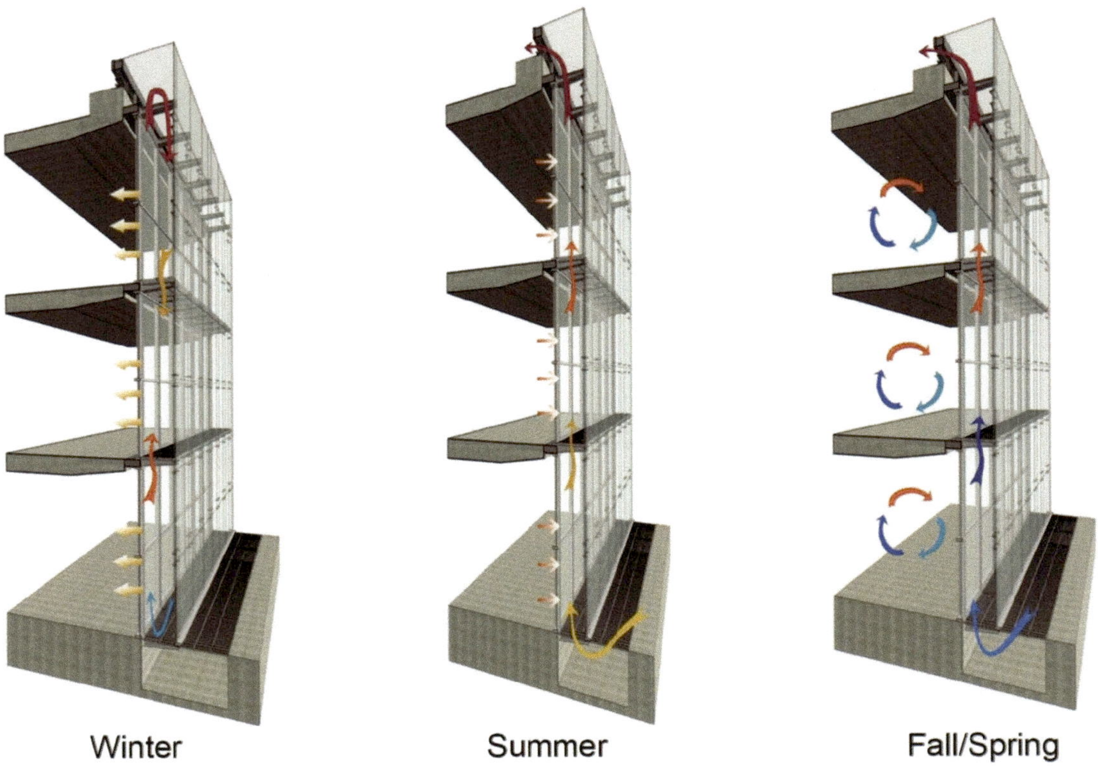

Figure 6.85 Façade Performance – Shoulder Seasons, Winter and Summer

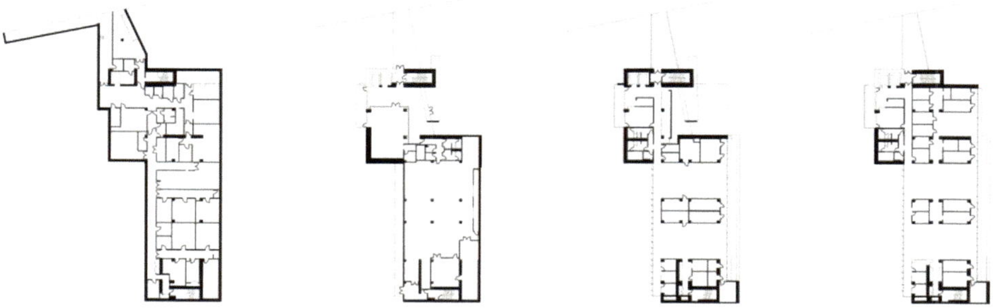

Figure 6.86 Plan Drawings

facility. The double skin façade designed by ZGF Architects LLP was selected because the project demanded a cutting-edge façade. The outer layer of glass is cabled-suspended from overhead cantilever building structure. T-shape outriggers at each floor level (to minimize cable spans) support the cables. The outriggers double as support for a steel maintenance grate that is located in the naturally ventilated cavity. The double skin façade takes advantage of the stack effect to exhaust warm air through a series of automated, louvered vents at the parapet that allows for control over airflow and optimize temperature. The southeast double skin façade has an alternating pattern of full-translucent frit and exterior laminated glass, creating a completely translucent effect to the façade while permitting transparent views from within.

In its pursuit of LEED Gold rating, passive strategies were included in the design. With the intent of mitigating harsh solar exposure, the shape and orientations of the plan, along with the use of concrete walls and stone cladding to provide thermal mass, help moderate temperature fluctuations. The double skin façade in warm climates provides benefits with proper ventilation. High-performance glass and air gap provide a thermal buffer against the outside temperatures. The double skin façade and a narrow floor plate help to reduce electricity consumption for artificial lighting by maximizing daylight throughout the interior by providing 92% of the interior lighting needs per LEED requirements.

CONCLUSION

The original concept used a shared cable truss to support the interior and exterior walls by drilled point supports (glass bolts) on both sides. This concept was ultimately scaled back to separate structural systems due to cost. The final design includes an interior skin with low-E coated insulated glazing supported by a unitized aluminum curtain wall system. The exterior laminated glass is cable-suspended from the overhead cantilever building structure. The vertical cables are supported by T-shaped outriggers at each floor level to minimize cable span and deflections under loading. These steel outriggers double as support for maintenance grating and shading devices located within the cavity. Point-fixed, bolted, laminated glass attaches on a 5'3" vertical grid to the cables.

Bibliography

Vaglio, J., & Patterson, M. (2010). Seeing double: Double-skin façade designs on the rise in North America. *Architects' Guide to Glass Magazine.*

Vaglio, J., Patterson, M., & Hooper, S. (2010). Emerging applications and trends of double-skin facades. *Enclos.*

ZGF Architects LLP Portfolio, Eli and Edythe Broad CIRM Center for Regenerative Medicine and Stem Cell Research at USC. https://www.zgf.com/project/usc-broad/

Figure 6.87 The Occidental Chemical Center Is the First Building in America to Incorporate a Double Skin Façade

Credit: Terri Meyer Boake

Occidental Chemical Center

LOCATION
Niagara Falls, NY

BUILDING PROGRAM
Office and Retail

ARCHITECT
Cannon Design Inc.

SIZE
200,000 SF

COMPLETION DATE
1980

BACKGROUND AND CONTEXT

The Occidental Chemical Center, built in 1980 by Canon Design Inc., is an early example of double skin façades. First designed as headquarters for the Occidental Chemical company, today, this building has a rocky past with many owners and mishaps during renovation. Like its original design, it still contains office spaces, but today it is occupied by other companies, stores, and restaurants. Located in Niagara Falls, New York, just near the Rainbow Bridge Toll Gate that

Figure 6.88 Building on Site

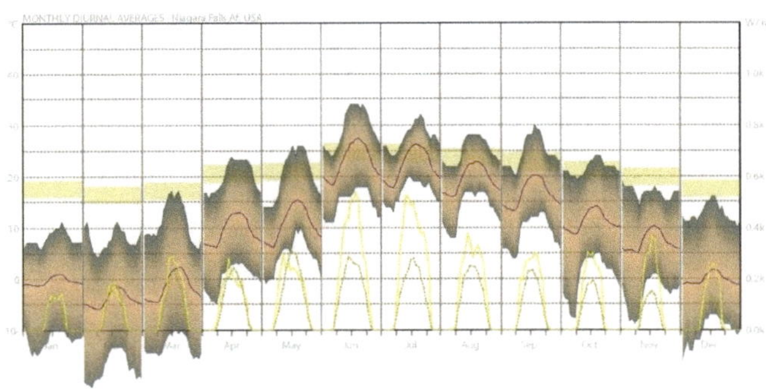

Figure 6.89 Hourly Temperature

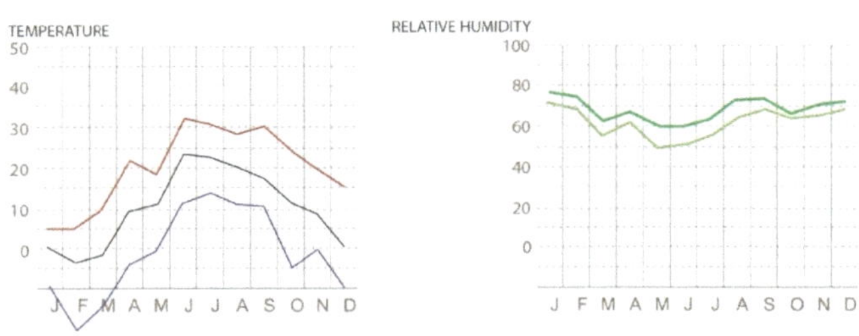

Figure 6.90 Climatic Data – Monthly Temperature, Relative Humidity

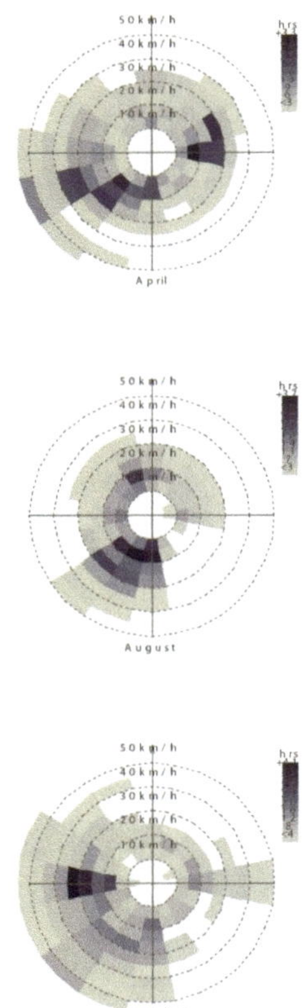

Figure 6.91 Prevailing Winds

separates Canada and the United States, this building had been a prime example of what the pros and cons of double façades could be.

Located in Niagara Falls, New York, the Occidental Chemical Center is the first building in North America to incorporate a double skin façade in its construction. Designed by Cannon Design Inc. in 1980, the building was originally intended to serve as the Hooker Chemicals & Plastics Corporation headquarters. [1] Hooker Chemicals & Plastics Corporation was purchased by Occidental Petroleum Corporation in 1968, although the name of the company has been altered multiple times since. Occidental Chemical Corporation, or OXY, is the company's primary Niagara Falls location, and the chemical plant is located in an alternate location near the Niagara River. The Occidental Chemical Center has served a multitude of hosts since its construction, including office, retail, and restaurant tenants. The building is currently home to the One Niagara Welcome Center, which offers sightseeing tours, two floors of shopping, an international food court, and a Native American Dance Theatre. [2]

Recent news in August 2021 reported that the nearby site of Occidental Chemical Corporation in Niagara Falls is set to close. [3] This announcement concludes decades of controversy involving Occidental Chemical's influence in the region. The then Hooker Chemicals & Plastic Corporation was blamed for America's worst environmental disasters involving chemicals. The nearby site of an abandoned canal became a dumping ground for chemical waste in the 1940s and 1950s, which resulted in a 1978 lawsuit of over 100 million dollars. [4] In contrast to the corporation's history, the Occidental Chemical Center has been regarded as one of the most energy-efficient buildings ever to be constructed and has won numerous awards on energy consumption, including the AIA and Illuminating Engineering Society, Energy Conservation Award from the Owens Corning Fiberglas Corporation, and the EPA Energy Star. [1]

DESIGN INTENT AND BUILDING FEATURES

Built shortly after the Niagara Falls environmental lawsuit and the 1970s oil crisis, Hooker Chemicals & Plastic Corporation desired an energy-conscious and cutting-edge headquarters. The building form was designed to incorporate views of the surrounding Gorge, with floor-to-ceiling glazing and a central core. The relatively typical structure consists of concrete slabs on all nine floors and a steel frame structure. To meet both the environmental and formal requests from the client, Cannon Design created a façade with two individual layers separated by a continuous air space, a technique that had not been utilized in America at the time of design. The air cavity acts as a buffer between the two layers of glazing, stabilizing the interior layer of glass. [Figure 1] Air is taken into the gravity through grilles at the base of the structure and released at the top of the structure. [Figure 2] In winter months, the cavity can be closed off when cold air circulation in the cavity would otherwise inhibit the stabilized temperature inside the building.

Figure 6.92 The Air Cavity Is Surrounded by Glazing on Each Side. Vents that Allow for Air Circulation in the Cavity Can Be Seen on the Ground and Floor Slab Above

Credit: Terri Meyer Boake

Figure 6.93 The Exterior Vents Draw Air into the Cavity at the Ground Level on the Structure's Exterior

Credit: Terri Meyer Boake

The façade construction consists of two green-tinted insulated glass layers, an air cavity, a louver system, and an additional single layer of clear glazing on the interior. [1] The louver system has significantly deteriorated since the building's construction. Today, many have been broken and removed. However, these horizontal louvers were once operable throughout the entire building and helped control the amount of light permitted to enter the interior of the building. The louver system worked with an intelligent system that senses the time of day and adjusts the louver direction accordingly to maximize comfort. When solar cells in the back of the louvers came into contact with light, the louver system would rotate to direct sunlight towards the ceiling. [5] The solar heat gain that would be picked up by the louvers could be exhausted through the cavity. In the winter months, this system was able to close entirely and act as an additional buffer layer. Just as the louver system was not controlled by individuals inside the building, the HVAC system was designed with similar intent. The envelope contains its air handling system, which occupants cannot adjust. This further aids in the building's energy savings

Figure 6.94 Remaining Louvers in the Structure Control Light Exposure to the Interior Spaces, but Are No Longer Operable

Credit: Terri Meyer Boake

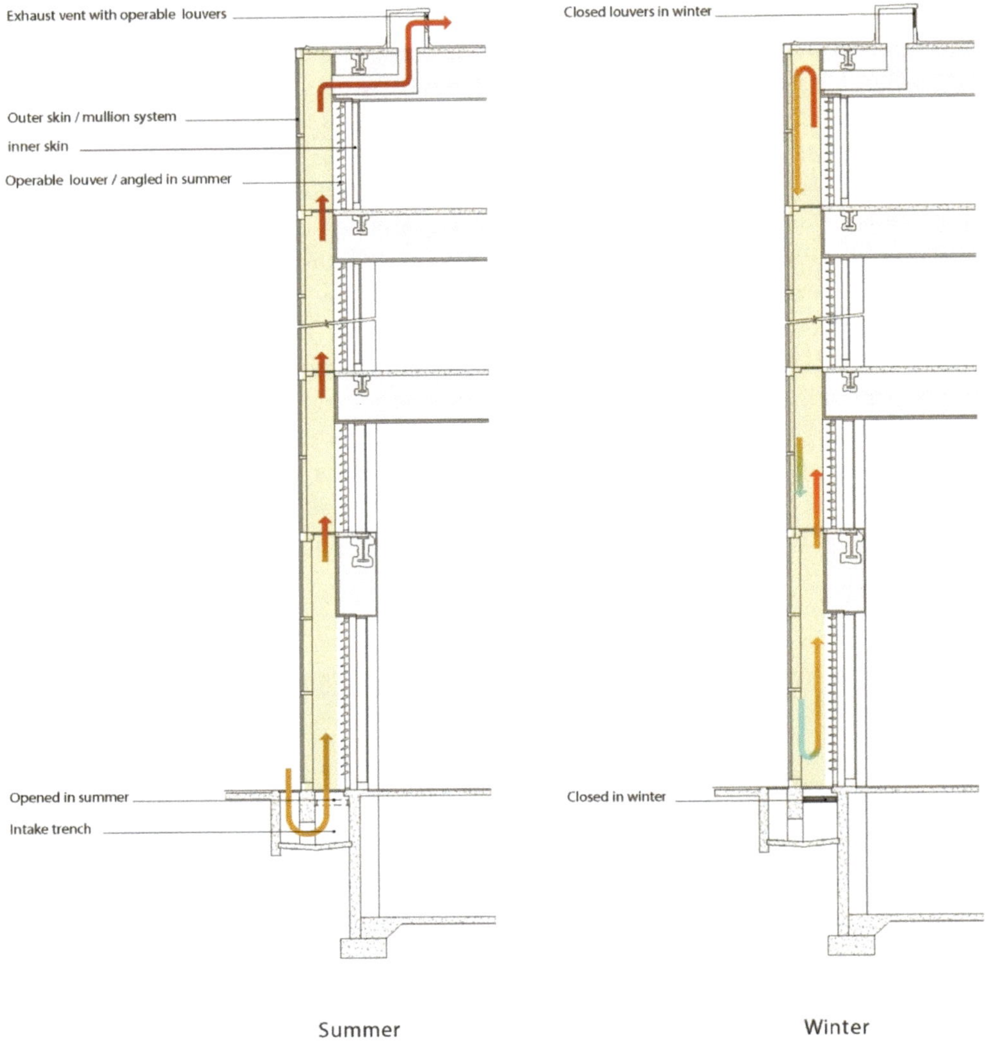

Figure 6.95 Wall Section

and works to maximize efficiency based on the time of year and amount of sunlight entering the building.

The configuration and energy saving intent of this DSF design is similar to that of the Seattle Justice Center, but the cavity is used to preheat air for the HVAC system during the heating season. As in the Justice Center, there are no operable windows.

In 2001, Terri Meyer Boake and Kate Harrison of the University of Waterloo studied this building to assess the DSF performance. The cavity was quite dirty due to a lack of maintenance. The automatically controlled louvers had not been

The wall section shows how the double skin allows for air circulation that can be closed off in winter.

The cavity between the layers of glass also contains a shading device made of horizontal louvers that can be tilted to control the amount of light entering the building, much in the same way as blinds. Unfortunately, this louver system is now broken, and many of the louvers have been removed.

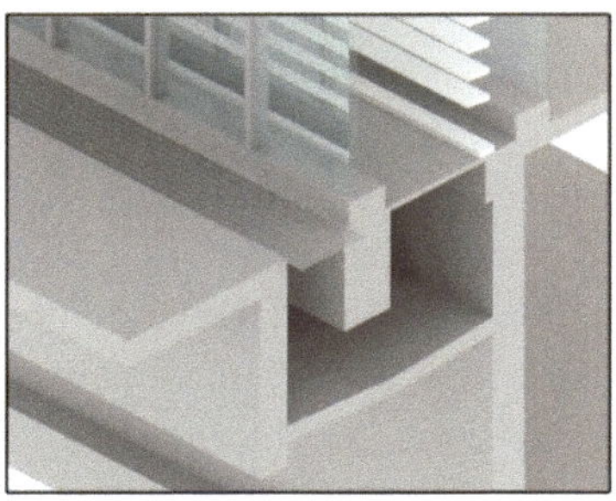

Bottom Grill and Outboard glass connection detail.

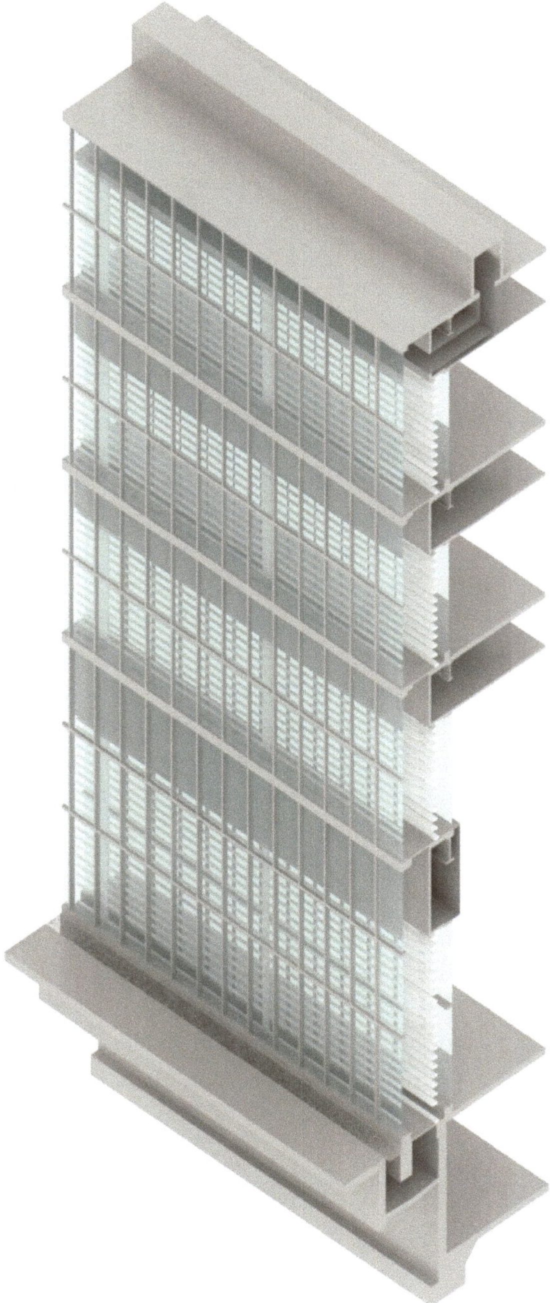

Figure 6.96 3D Wall Section

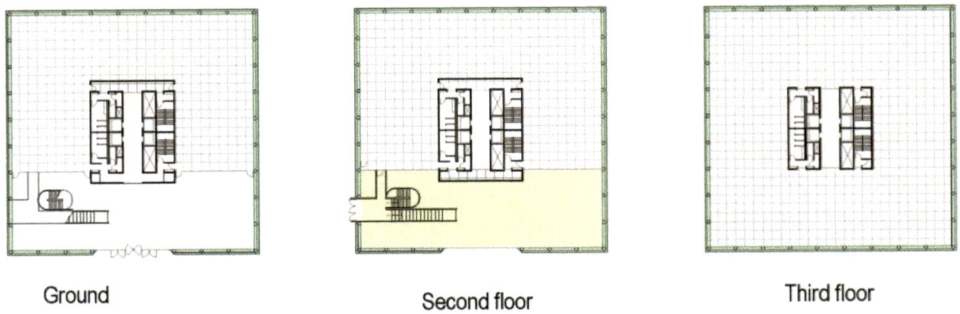

Figure 6.97 Plan Drawings

functioning for about four years, and the motorized dampers were covered with plywood. There was no airflow through the cavity, which caused it to overheat; occupants complained of discomfort due to the malfunctioning system. The faults were found to be primarily a result of a failed control system.

The wall section shows how the double skin allows for air circulation that can be closed off in winter. The cavity between the layers of glass also contains a shading device made of horizontal louvers that can be tilted to control the amount of light entering the building, much in the same way as blinds. Unfortunately, this louver system is now broken, and many of the louvers have been removed.

Each floor features a drop ceiling that hides the steel structure and ventilation systems.

BUILDING PERFORMANCE ANALYSIS

The original design of the Occidental Chemical Center estimated that the building would consume only one-third of the energy an average building without a double skin façade of the same size would consume. In addition to the envelope, the building also makes use of its location in Niagara Falls and uses hydroelectric energy. However, changes in program and materials as construction finished altered the performance of the structure significantly. A last-minute change in the program for office spaces to include research space resulted in gas-fired boilers that remained obsolete and were never used to heat the building. [1] A glare resulting from a reflecting coating on the louvers required repainting and lower efficiency than originally intended.

Unfortunately, the building has experienced much deterioration and a remarkable decrease in efficiency since its construction compared to what was originally intended. With the breaking and removal of much of the louver system, current tenants have resorted to installing standard vertical blinds controlled by the occupant. The remaining louvers are situated in a fixed horizontal position and block only 50% of the solar radiation on the south façade. [1] Current renovation at the east entrance has forced

Figure 6.98 Highlighted Red Panels Display Clouding on the Glazing System Due to Years of Deterioration and Lack of Repairs

Credit: Terri Meyer Boake

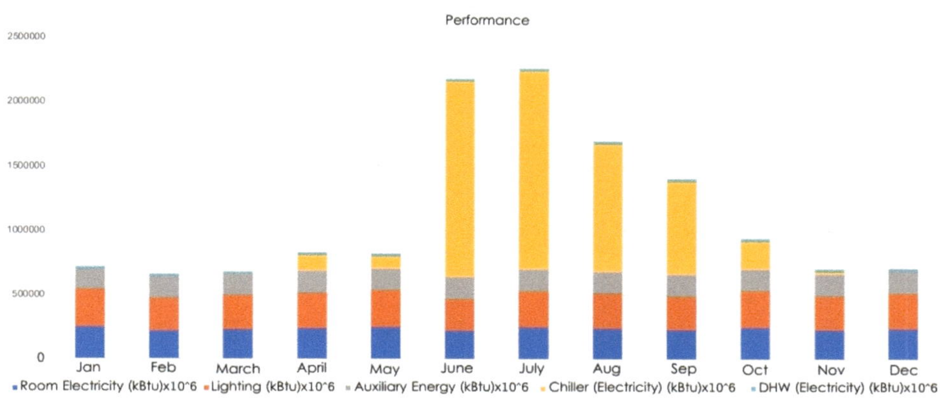

Figure 6.99 Performance Analysis of the DSF (kBtu)

the exterior intake grilles to be covered with plywood to stop dust and dirt from accumulating in the cavity. Years of disrepair and neglect have additionally left much of the glazing clouded and dirty [Figure 6.98].

CONCLUSION

The design was intended to be the first of its kind, incorporating the interior and exterior environments to create comfortable spaces and reduce energy costs. Overall, the building did not accomplish its goals due to the state of disrepair that it quickly fell into. Had it been well maintained the building would have easily been one of the most cost efficient and environmentally friendly buildings of the early double façade era.

The Occidental Chemical Center was designed with a focus on energy conservation and the inclusion of an innovative double skin façade. However, poor maintenance and lack of attention to future problems resulted in a system that quickly deteriorated and failed to accomplish its energy conservation goals. The current system is unable to accurately protect against solar radiation and circulate fresh air for maximum comfort. As a result, the building currently functions as an ordinary building with conventional equipment. It is important to not only recognize the Occidental Chemical Center's innovation as America's first double skin façade, but to also understand and learn how the system failed so that other structures will not face the same challenges and unfortunate outcome.

References

1 Meyer Boake, T., Harrison, K., & Chatham, A. (2002). The Tectonics of the Double Skin: Green Building or Just more Hi-Tech Hi-Jinx? https://www.tboake.com/pdf/tectcase.pdf

2 "One Niagara Welcome Center." Niagara Falls USA. Accessed September 28, 2021. https://www.niagarafallsusa.com/directory/one-niagara-welcome-center-shopping/.

3 Specht, C., Mickey, S., & Wood, L. (2021). Occidental Chemical's Niagara Falls Plant to Close, 150 People Impacted. WKBW, August 19,. https://www.wkbw.com/news/local-news/occidental-chemicals-niagara-falls-plant-to-close-150-people-impacted.

4 Environmental Protection Agency (EPA) (2016). U.S. Sues Hooker Chemical at Niagara Falls, New York. August 16. https://archive.epa.gov/epa/aboutepa/us-sues-hooker-chemical-niagara-falls-new-york.html.

5 Meyer Boake, T., Chatham, A., Harrison, K., & Lum, K. (2003) Doubling up II. *Canadian Architect*, July 31. https://www.canadianarchitect.com/doubling-up-ii/.

Figure 6.100 Manitoba Hydro Place Tower

Manitoba Hydro Place Tower

LOCATION
Winnipeg, Canada

BUILDING PROGRAM
Office

ARCHITECT
KPMB Architects

SIZE
695241 SF

COMPLETION DATE
2009

BACKGROUND AND CONTEXT

In sustainable skyscrapers, priority has been given to passive systems of heating and cooling over active and mixed-mode systems because they consume the least energy and from renewable resources. Manitoba Hydro Place is an 18-story office building, which features a "bioclimatic" design adapted for the Canadian climate.

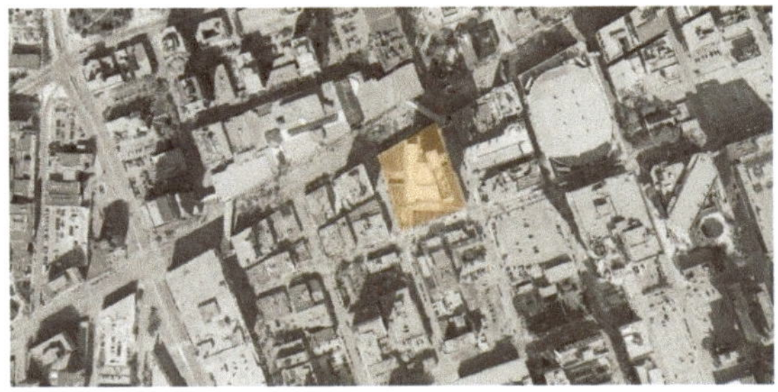

Figure 6.101 Building on the Site

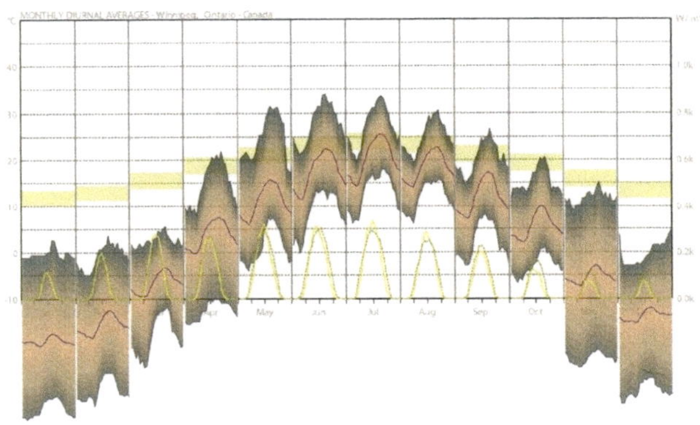

Figure 6.102 Hourly Temperature

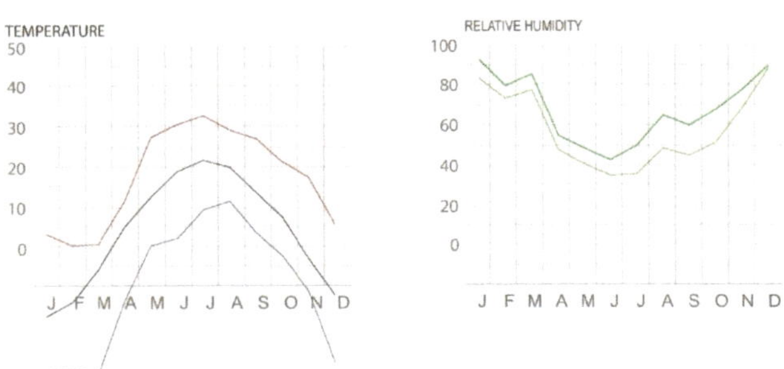

Figure 6.103 Climatic Data – Monthly Temperature, Relative Humidity

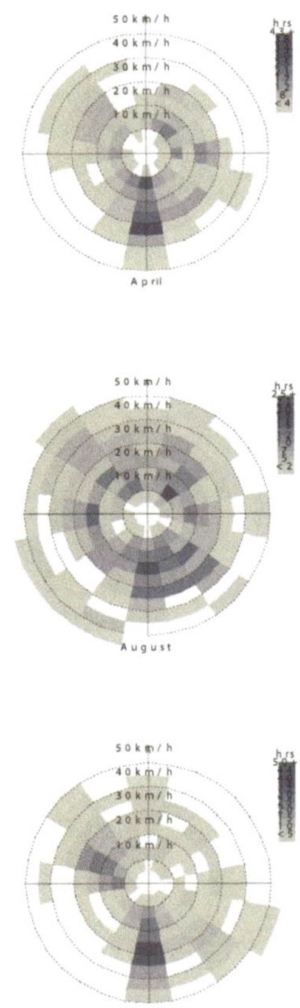

Figure 6.104 Prevailing Winds

Manitoba Hydro was completed in 2009. It is the primary energy utility in the province of Manitoba, the fourth largest energy utility in Canada and it offers one of the lowest electricity rates in the world.

The site was specifically chosen because of its closeness to public transportation routes and the untapped opportunities to design a building to harvest passive solar heating and daylighting. The orientation and massing strategies were carefully analyzed to optimize the potentials of the site.

Manitoba Hydro is located in Winnipeg, Manitoba, which is one of the coldest cities in the world. The climate and temperature of Winnipeg is extreme and fluctuates dramatically over the year. Winter temperatures average around -35 °C, climbing to +34°C by the summer time. The severe climate of Winnipeg has 8.9 m/s winds in the wintertime but also it is very sunny during the cold season that results in an opportunity to incorporate passive solar heating and daylighting which reduces the lighting loads and creates a hybrid ventilation.

DESIGN INTENT AND BUILDING FEATURES

Manitoba Hydro's form consists of two 18-story twin office towers resting on a stepped 3-story public arcade that includes commercial space. The towers converge on the north side and open to the south for capturing maximum sunlight and winds. The towers with narrow floor plates and floor-to-ceiling glazing let the daylight enter the space deep. Every six floors of the towers open to south-facing winter gardens with waterfalls to humidify and dehumidify the air entering the offices. A 115-meter-tall solar chimney is located on the north side of the building. It rises above the towers and is a key element for the passive ventilation system. It works with winter gardens that extend through a number of floors with mechanical louvers at each end of the atrium enabling the space to be closed or opened according to weather conditions ventilating the air through the stack effect in the solar chimney.

The solar chimney, during the summer months and shoulder season, draws the exhausted air out of the building. In the winter months, the heat recovered at the chimney is brought to the bottom to preheat the air inlets.

The building consumes 80 kWh/sq.m per year which is 66% more efficient than the model national energy code for buildings. Beyond being a high-performance building, it also provides 100 percent fresh air, which results in a healthy environment. The building uses a holistic approach to be a Gold LEED rating building. The sustainable features to achieve the results are:

- Passive strategies: Thermal mass, Daylighting, Geothermal, Natural ventilation: Solar Chimney, Winter gardens, Water features.
- Mechanical strategies: Radiant heating and cooling, Displacement ventilation, and Double skin Façades.

Figure 6.105 Plan Diagram

Figure 6.106 Building on the Site

Credit: Terri Meyer Boake

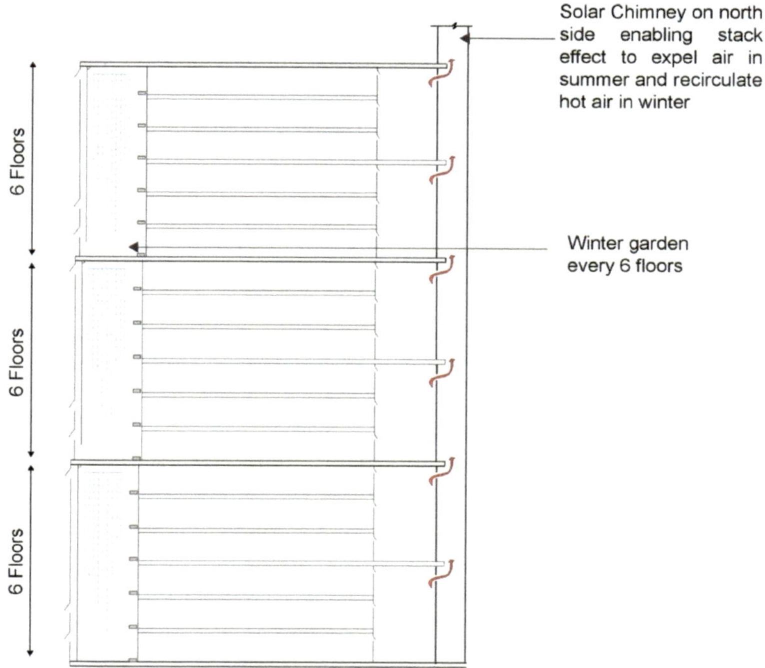

Figure 6.107 Section Diagram

Daylighting

Majority of energy consumption of the building goes into the lighting system. Providing light not just as a means of energy efficiency but also health and well-being of the occupants, provided by narrow floor plates of the towers.

The two office towers are oriented towards the west and northeast. All floor-to-ceiling glazing and ceiling height of 3.3 meters and double skin façade provide daylighting with motorized windows and large automated louvers to minimize solar heat gain and glare. Louver blades act as a light shelf at the top and penetrate the light deeper inside the room by reflecting light onto the white ceiling. The floor plates are narrow so no employees are more than 9 meters from the window. The integrated shading systems are perforated to allow transparency and penetration of the daylighting on the east and west orientation while blocking the solar heat gain and glare.

Natural Ventilation

From the onset of planning, Manitoba Hydro is providing 100% fresh air in contrast to North American office buildings that recirculate the air. The building is oriented towards the south so the prevailing wind can naturally ventilate the structure. The building provides direct outside air regardless of the season as well as individual control for the occupants.

The interior of the curtain wall is manually operable and allows occupants to control their environment while the exterior has automated wall vents to adjust the temperature and airflow.

The natural ventilation strategy of the building is provided by a double skin façade system, which supplies air to the offices, and a series of six-story winter gardens. They draw the air and precondition it before it enters the office spaces through vents in the raised floor. The 115-meter-tall solar chimney on the north side exhausts the return air rises from the occupants and other sources in the offices. In the winter, the solar chimney draws exhaust air down to heat the parking and pre-heat incoming cold air from the winter gardens.

In the shoulder seasons, the building only provides outdoors fresh air through the use of automatic and manually operated windows.

Thermal Mass

The structure of the building is concrete and it was left exposed to create thermal mass to moderate the temperature fluctuations. The structure creates radiant slabs to exchange the heat with the offices.

Winter Gardens

On the north and south side of the winter gardens, there are interconnecting stairs that provide opportunities for physical activity and interactions among employees. The winter gardens act as the "lungs" of the building; they feature a 24-meter-tall water curtain that will

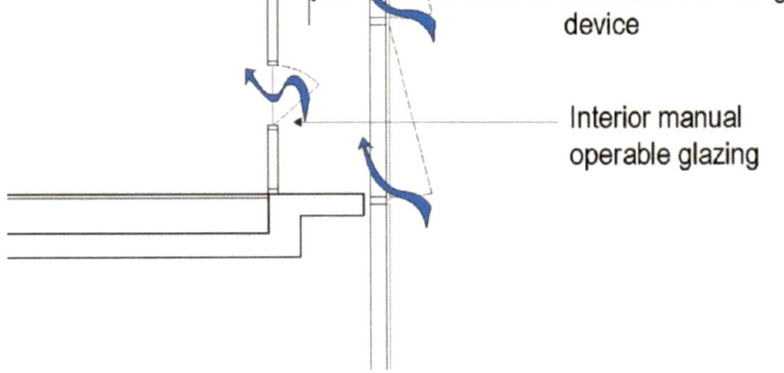

Figure 6.108 Natural Ventilation Strategy Atrium Section

Figure 6.109 Top of the Chimney

© Photo by Terri Meyer Boake

Figure 6.110 Natural Ventilation – Double Skin Façade

© Photo by Terri Meyer Boake

Figure 6.111 Exposed Concrete Slab to Create Radiant Floors

© Photo by Terri Meyer Boake

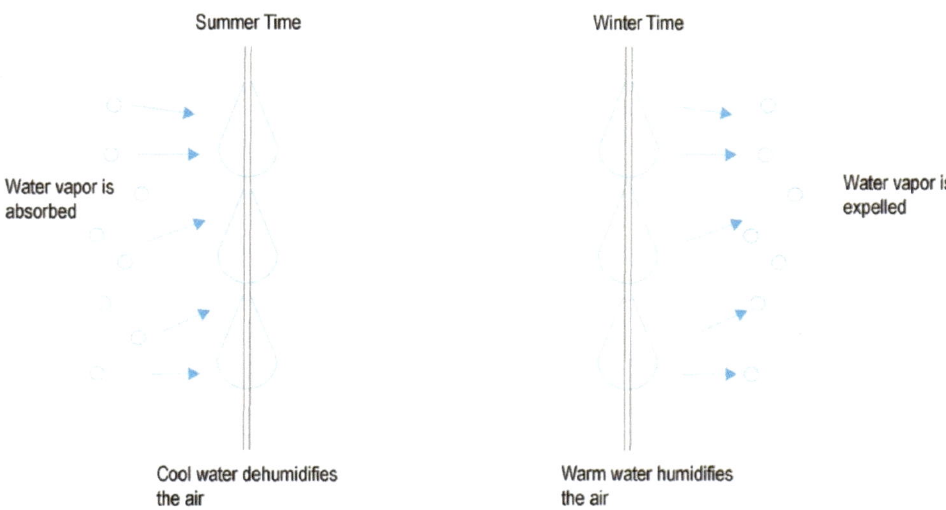

Figure 6.112 Water Flowing Down in the South Atrium

Figure 6.113 (a) Tall Curtain Wall (b) Water Ribbons (c) Winter Garden Atrium Space

© Photos by Terri Meyer Boake

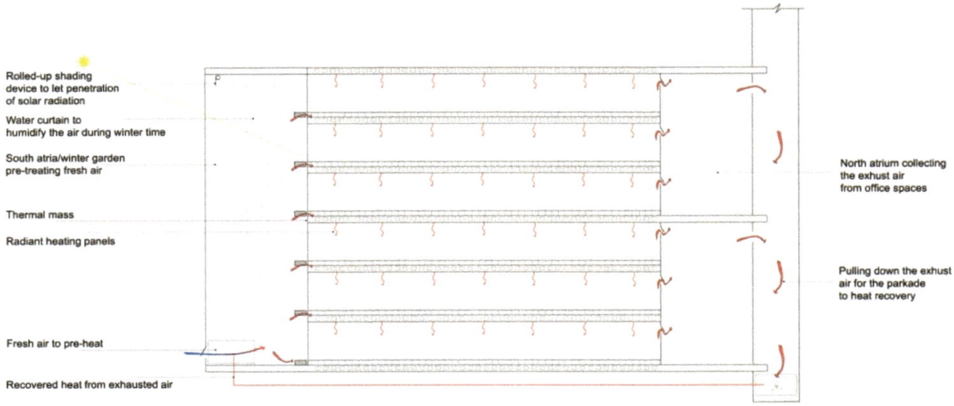

Figure 6.114 Heating System Diagram

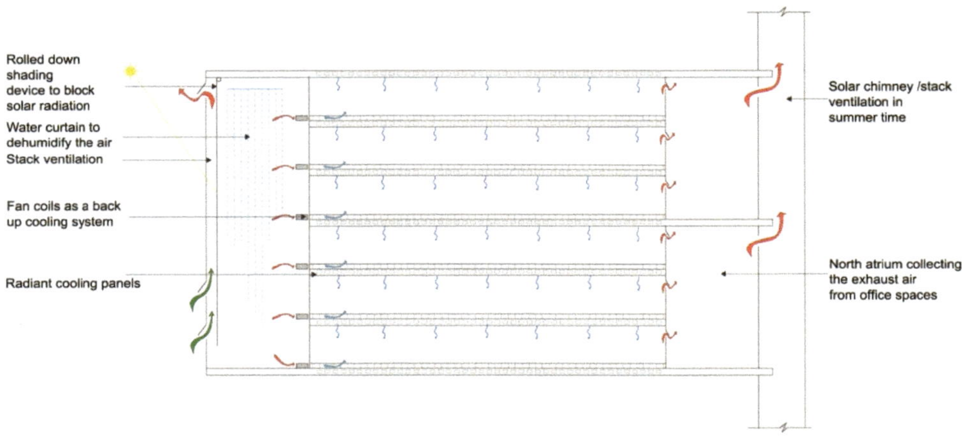

Figure 6.115 Cooling System Diagram

humidify the air during the winter month. The louvers in the south-facing double wall let the fresh air enter the winter garden and via the 280 tensioned mylar ribbons extended from ceiling to floor, the incoming air can be heated by the sun and humidified by the water. During the summertime, this water that goes down the ribbons is chill and can dehumidify the air that enters the atria.

Heating and Cooling System

One hundred percent of the cooling load and 50% of the heating load is covered by a geothermal heat pump system. The main heating and cooling source of the Manitoba Hydro building is exposed radiant concrete ceiling slabs. The water pumped in the tubes that are embedded in the concrete floor slab heats and cools the spaces by thermal radiation. This

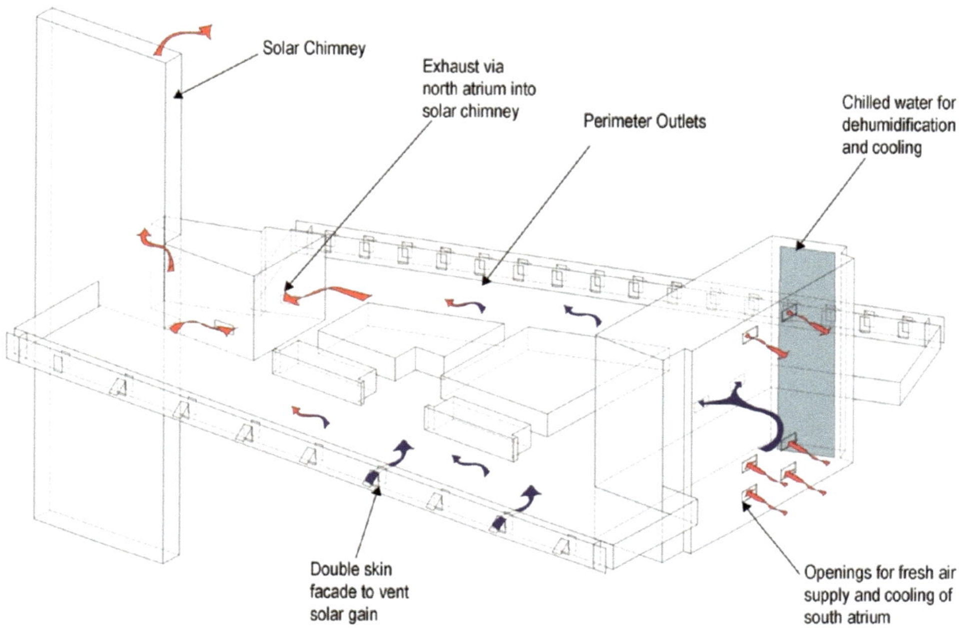

Figure 6.116 Airflow Concept for Summer

water system provides better comfort and uses less energy in comparison with air heating systems. The radiant heating and cooling slabs are located in the office space. However, the excess energy can run through the Watergardens and the buffer zone in the double skin façade to maintain a temperature and reduce the heat loss

Displacement Ventilation

Displacement ventilation introduces 100% fresh air into office space through a raised floor system at a very low velocity. Air enters at the floor level and moves gradually as it is warmed by occupants and rises. This exhaust air exits the solar chimney on the north side. This system reduces energy as it just needs to provide enough fresh air for the occupants and eliminate the conventional ducts under the ceiling.

Geothermal System

There is a closed-loop system consisting of 280 boreholes, six inches in diameter, 400 feet deep, spread between the building's foundation piles and caissons. Each of these boreholes contains glycol that extracts the heat from the building and returns it to the ground during the summertime and the flow is reversed during the winter season. Conditioned water is circulated in tubes in radiant cooling slabs and provides 100% of the temperature conditioning. Through this radiant

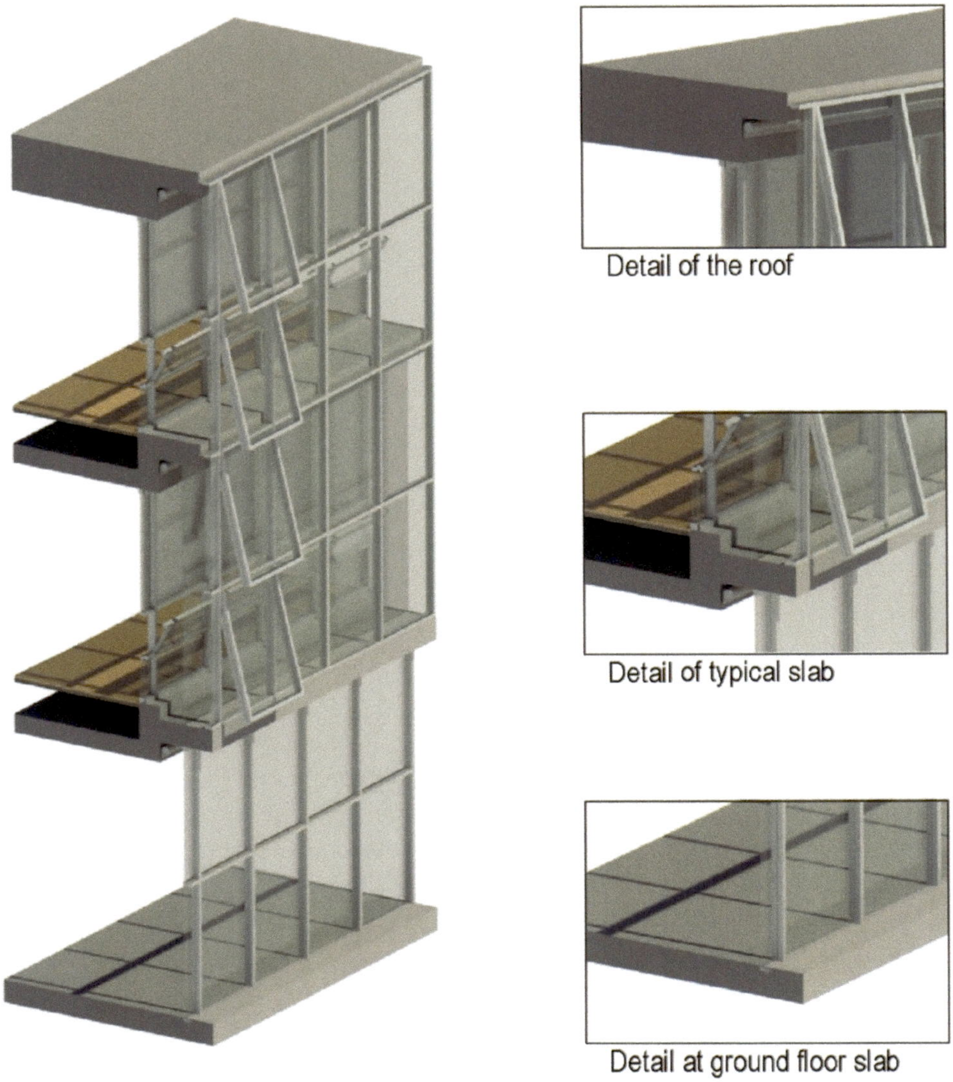

Figure 6.117 Double Skin Wall Section in 3D

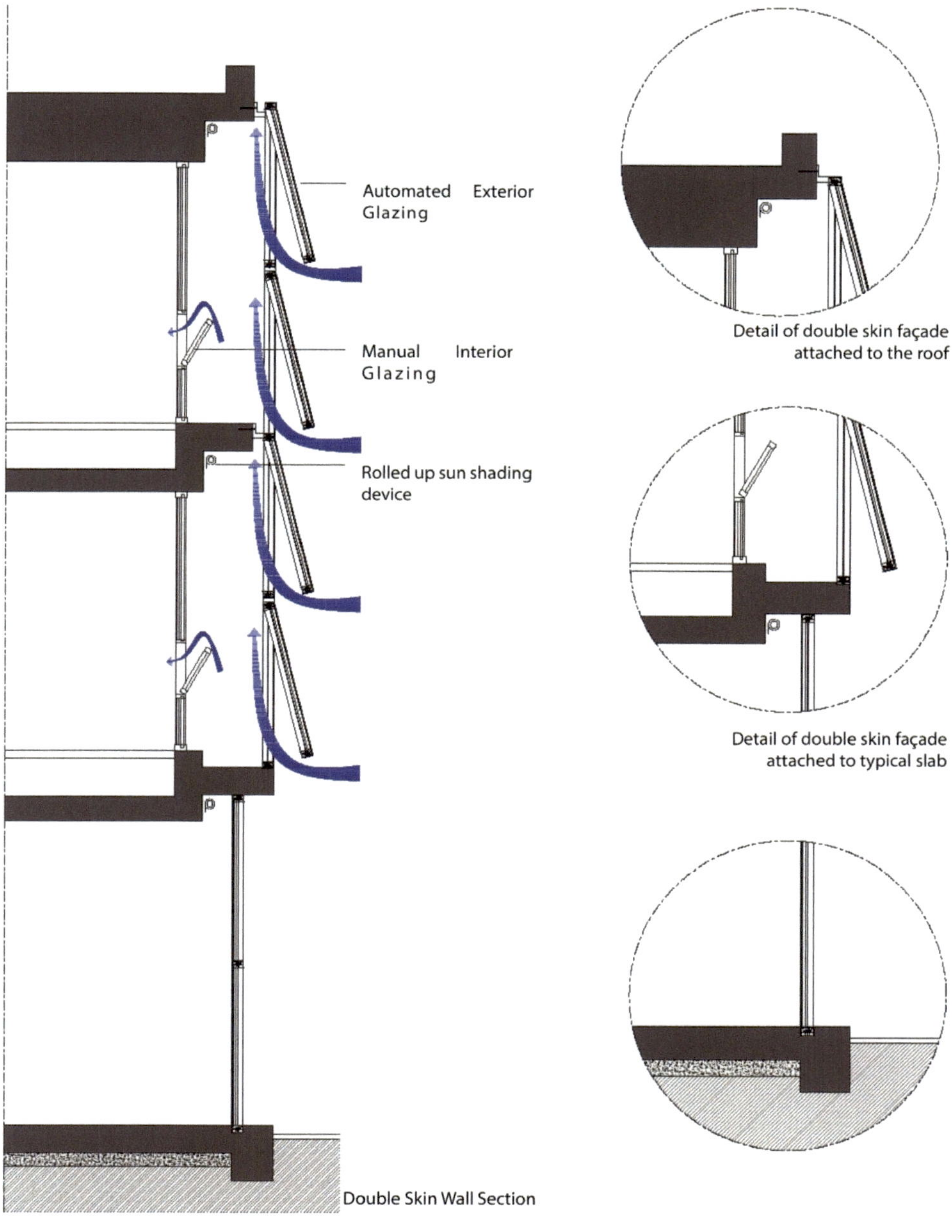

Figure 6.118 Double Skin Wall Section

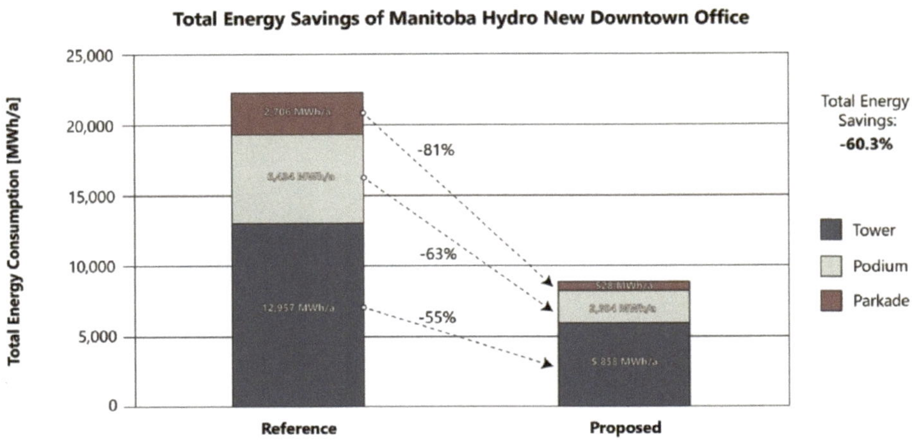

Figure 6.119 Performative Analysis by KPMB

Figure 6.120 Energy Gradient Maps of the Interior Spaces of Manitoba Hydro Place

Credit: Terri Meyer Boake

heating system, the geothermal installation provides approximately 60% of the heating in the radiant slabs.

BUILDING PERFORMANCE ANALYSIS

The energy consumption analysis of the building was compared to a similar energy benchmark. This comparison shows that the building's projected performance surpasses those benchmarks. The total annual thermal energy consumption ranges between 105 and 128 kWh/sq.m. It is important to mention that these figures fall below all benchmarks including fully air-conditioned offices (190 kWh/m2), offices mechanically ventilated and heated but not cooled (160 kWh/m2), and most surprisingly naturally ventilated office buildings (135 kWh/m2) [1].

Two years of optimization of energy performance led to an unexpected result considering the extreme climatic conditions. All the optimization strategies resulted in the reduction of energy loads. The first-year operation cost was 161 kWh/m2 and that was reduced to 85 going beyond the target of 120. Which in comparison with the benchmarks is 70% better.

CONCLUSION

Manitoba Hydro Place incorporates both passive strategies as well as mechanical strategies for the heating and cooling of the building. It touches upon all strategies that can be used in the climate of Canada.

Figure 6.121 Exterior Space

Photo by Terri Meyer Boake

It successfully brings in natural ventilation in an extremely cold climate. The use of the water curtain in the winter gardens to humidify and dehumidify the air is efficient and can be used in the summers and winters. The south atrium helps in solar gain during winter and brings in fresh air during summers. The north atrium expels air out to the chimney during summer. In winter, this hot air is recirculated within the building as preheated air. The combination of the automated and manually operable glazing on the west and north-east façade gives better control of the temperature for individual users. The use of a shading device and radiant floor also helps in controlling the temperature of the building during summers and winters. Exposed concrete is used to increase the thermal radiation of the façades while low-iron glass is used to increase the solar gain. Apart from these aforementioned strategies, geothermal energy is also used to heat the building during extreme winters. Overall, the Manitoba Hydro Place tower uses every strategy possible to reduce energy consumption and is successful in reducing the energy to almost 60% than an average skyscraper building in North America.

References

1 Gonçalves, J. (2010) *The environmental performance of tall buildings.* Earthscan.

Bibliography

Akerstreem, T. Knirsch, A. & Pauls, M. (2013). Manitoba Hydro Place: Energy Efficiency 2.0. ASHRAE Transactions, 119, F1–F8.

CANADA. DOWNTOWN WINNIPEG BUSINESS IMPROVEMENT ZONE. (2010–2012) *Downtown Trends, Downtown Winnipeg Market Research.* [Online] Available from: http://www.centreventure.com/PDF/Downtown Trends.pdf. [Accessed: 9 July 2013]

Manitoba Hydro Place. American Institute of Architects. [www.aiatopten.org/node/110].

Manitoba Hydro Place. Building Facts. [www.hydro.mb.ca/corporate/mhplace/quick_facts_building].

Manitoba Hydro/KPMB Architects. (24 December, 2009). *ArchDaily.*

CHAPTER 7

Façade of the Future

The increasing growth of chronic diseases and epidemic viruses urges the need for innovative smart building technologies to help with preventing, mitigating, and controlling the spread of highly contagious diseases. Currently, people spend more than 90% of their time indoors. The rapid spread of COVID-19 further underscores the role of buildings and façades in providing occupants with a healthy environment.

The next generation façades must be based on environmental principles that also consider occupants' health and comfort. Since buildings have highly complex design details, every component of these details can lead to a complex interaction of the façade with the environment.

Conventional façades have a limited extent to which energy can be saved. They cannot take advantage of desirable outdoor conditions to provide comfort for indoor users. The future façade needs to respond to the external climatic factors while considering indoor user comfort. With all the seasonal variations, shifting weather patterns, and indoor comfort metrics, there is a need to make façades more responsive. Since the climatic factors are dynamic, building façades should respond to the corresponding fluctuations. The façades need different typologies to enhance the control and modify the winter losses or summer solar gains based on external forces and the desired internal comfort. [1]

Over the past decade, modern advancements in façade systems have led to high-energy performative buildings. Future façades are different from conventional ones in that they can have characteristics defined in a domain which can then mediate and change between the inputs of the environment. [6] Dynamic façade technologies are the components of advanced fenestration systems that allow occupants to interact and regulate the environment based on their comfort and preferences while reducing energy consumption. The significant difference between the Dynamic

DOI: 10.4324/9781315718750-7

Façade System and the fixed one is the ability to change according to external and internal factors to maintain occupants' comfort and needs. On the other hand, the fixed façade is designed primarily to address the climatic conditions without being responsive to any changes. In contrast, the dynamic façade is categorized based on the type of morphology (blind, louvers, dynamic egg-crate, etc.), type of actuation (manual or automated), mode of actuation (intrinsic material properties, remote control, environmental sensing, occupant-cantered), and interfaces that could adopt to control and manage lighting, solar insolation and thermal radiation on a building façade.

The sun is a free energy source, and façades should take advantage of this for lightning, heating, and cooling. Wind and rainwater are also free resources. Façades should collect rainwater and encourage the wind to enter the building for ventilation. When a building is too warm, the façade should block the sun. When more lighting is needed, the façade should open to allow more natural light. Next-generation façades should also enable the building to breathe. These will be biophilic, allowing a building to take advantage of outside air and ventilation processes. Whether the façade features actual vegetation or mimics the act of breathing, the future of façades relies on the fact that they can take advantage of exterior air and introduce ways in which external air can be brought into the building by way of sources other than HVAC systems. On top of this, next-level façades should adapt to exterior air movement and all exterior influences on the building. This should be done so that building occupants are comfortable, promote sustainability and aid in optimizing the building's energy consumption.

Architects have always imagined the potential of future buildings that can be similar to the human skin's adaptive response to changing environmental factors. [2] Through advances in material technology and automation in building systems, the road is being paved between the adoptive envelopes and the intelligent responses of the user's behavior and skin to external stimuli. [3]

There are many ways for a façade to meet all these needs through different modes of movement. Some buildings may favor certain areas of sustainability over others, or if a site has a greater potential to offer external influence through specific characteristics. A façade may be geared towards taking full advantage of sunlight in the Midwest or benefiting from air movement on coastal land. Images 7.1–7.7 show how façade movement can be applied to the adaptation of façades to react to changing environments.

For instance, one responsive behavior in the dynamic façade is contingent on external and internal forces. An automated system modifies the different glazing types based on the desired lighting. [1]

By employing adaptive façades, a computational optimization is presented to enhance the energy efficiency of the building, one that can define an adoptive consciousness for the façades to interact with various external factors. [4]

Since the design of the responsive façades is an optimization matter, the traditional approaches with only a single design solution, which cannot be adjusted based on the varying environment, fail to address the problem. [4]

To a certain degree, buildings are being constructed based on modularity and standardization. Most of the flexibility lies outside the building, while the core and the exterior shell remain excluded.

If buildings can be repurposed with a change in function, can next-level façades adjust to meet future needs? What if these elements can be flexible enough to change in consort with the interior? Façades are starting to incorporate these principles, but the next generation of façades will likely satisfy these requirements. They will adapt to the changes we make within buildings themselves. Figure 7.1 shows a building with movable pallets on the façade.

Although the materials may not be efficient to use on a façade, the principle of adaptability is present. The façade of the future will be the next level of this concept. A façade that can adapt, using the latest technologies and materials. Panels that can move or slide to meet changing needs within the building.

One way to address the responsiveness is to employ the systems that have active components, which perform through the movable parts in a system, for example. This intelligent kinetic shading system can change its aperture angle through a sensor-based control system. [5]

Another aspect of this issue is the focus on using responsive materials whose physical features are changeable based on environmental parameters. [3]

Figure 7.1 Wooden Pallet Building by Malka Architecture

Designing the future façade lies in its adaptivity. The use of this feature depends on a sequence of time-varying stages based on environmental factors. It needs to adapt with flexibility to many scenarios and consider short-term changes. [3–6]

The façade in Figure 7.2 introduces the movement of different height tubes spinning in response to sunlight. As more sunlight is desired, the façade elements aim to open the curtain wall and allow natural light.

The façade can also introduce vegetation as part of its form to symbolize sustainability and allow the façade to be a living element.

Façades of the future will be more than artificial objects but will feature natural elements as well that can be used to reduce carbon emissions and lower the embodied energy needed to manufacture a typical façade system. Façades of the future feature natural elements as well, ones that can be used to reduce carbon emissions and lower the embodied energy needed to produce a typical façade system.

The images above represent the ways façades can move to open or close in order to allow or block natural elements. The purpose of these two specific façades is to control the natural light in the space. This opens a wide variety of possibilities that employ different shapes, methods and technicalities of façades to accomplish similar goals. Façades of this nature

Figure 7.2 The Bund Finance Center Designed by Foster + Partners

Credit: GillyBerlin

can not only be used to control natural light but air movement as well.

Previous iterations of Architectural Dynamic Systems were focused on the function or the augmentation of the shape of space to increase efficiency, rather than on developing an improved indoor sensory environment (i.e., climate and lighting conditions). In contrast, the primary objective of a Dynamic Façade System is to provide a comfortable indoor environment and reduce building energy consumption. The aesthetic appearance of a façade is, of course, an essential aspect, but the aesthetic qualities of the kinetic system should, more importantly, be developed with consideration of its functional performance.

The difference between a standard façade system and a Dynamic Façade System is that the latter responds to external climate changes by using its control system. The primary mechanisms in kinetic systems that allow different movements are classified as follows: rotation, folding, translation and hybrid in an umbrella or folding and unfolding panel applications [7].

In order to design for kinetic movement, designers of a Dynamic Façade System must take into account several factors such as building program, orientation, site location, climate, occupant comfort, outside views, etc. Thus, it can be quite challenging to develop a Dynamic Façade System that satisfies all the possible factors and effectively determines appropriate kinetic movements. In the following sections, a few examples of kinetic façade systems explain how dynamic façades work and how the system contributes to reducing energy consumption and improving overall indoor comfort.

The Arab World Institute (AWI) in Paris, France (1987) by Jean Nouvel, is an excellent example of a dynamic brise soleil, which is able to regulate the amount of daylighting and solar heat penetration. The building consists of 11 floors and a total floor area of 181,850 SF. Based on the amount of sunshine received by the building (insolation), the kinetic brise soleil system was installed on the south façade in order to better regulate natural light.

Once installed, the apparatus is set to execute a maximum of 18 movements per day. The metal shutters of the façade system, with their varying opening sizes and kinetic mechanisms, are controlled by a centralized computer system, while the dynamic modules are situated between two glass panels to protect against weathering [8]. The façade, which faces southward, consists of 240 motor-controlled openings that filter light and are constructed of a metallic material. Initially controlled by photosensors, the wall is no longer operable.

The GSW Headquarters in Berlin, Germany by Sauerbruch Hutton is a 22-story office structure consisting of a total of 581,251 SF dedicated to office and retail space. The west façade boasts one of the primary features of this building, where the vertical louvers of each module provide rotating and sliding movements based on the user's preference or external weather conditions. The individual occupants can control all moveable elements of the façades; however, they are

Figure 7.3 Arab World Institute and Aperture Façade System (1987)

Figure 7.4 GSW Headquarters and Vertical Moveable Façade System (1999)

Credit: Seier+Seier

additionally operable by a central building management system. The vertical louver system plays an important role in reducing active heating and cooling. Furthermore, this façade also boasts a secondary glazed skin that assists with ventilation and cooling of the building. This double façade system serves as a second buffer for thermal control and acoustic attenuation. The deployment of this strategic façade design contributes to savings in energy consumption of up to 40% compared to traditional façade designs [9].

Kiefer Technic Showroom in Bad Gleichenberg, Austria by Ernst Giselbrecht + Partner has two stories for a hybrid exhibition and offers a total floor area of 5,866 SF. The kinetic façade system of this building can move according to the external weather conditions by using a foldable horizontal shading system. This kinetic façade consists of aluminium posts and transoms with protruding bridges for maintenance, coated with white plaster. This responsive façade system minimizes the cooling load by maintaining a constantly moving shield against external heat.

The Thyssen Krupp Q1 building in Essen, Germany, by JSWD Architekten of Cologne is a 10-story office building boasting a total floor area of 118,295 SF. The kinetic system used here consists of a vertical rotatable louver that controls the ratio of the opening. Furthermore, the rotational motion of this vertical louver is directly connected to the indoor illuminance sensor in order to optimize the daylight conditions of the indoor space [10]. Each vertical louver unit is composed of metal slats, similar to a venetian blind, to control daylight more efficiently. Close

Figure 7.5 Kiefer Technic Showroom and Horizontal Foldable Façade System (2006)

Figure 7.6 ThyssenKrupp Q1 and Rotational Vertical Façade System (2010)

Credit: Daniel Mennerich

to 400,000 stainless steel slats are used to blanket the entirety of the façade. The building deploys rotational movement of the façade components to ensure improved viewing conditions that are dependent on the external weather; while simultaneously maintaining the benefits of venetian blinds, including control of daylight and blockage of solar radiation. This kinetic louver system reduced the building energy use to less than 150kWh/m^2 per year in tandem with other sustainable systems [11].

The Façade as Powerplant

RMIT Design Hub designed by Sean Godsell Architects in association with Peddlethorp Architects has a translucent "smart skin" consisting of more than 16,000 sandblasted glass cells, some of which have photovoltaic collectors to harness solar power. The cells track the sun utilizing the building computer automation system to help shade and power the building. The inner skin is a high-performance, double-glazed layer that is designed to be upgraded over time to accommodate new solar technologies.

The outer skin of the Hub integrated automated sun shading that includes photovoltaic cells, evaporative cooling and fresh air intakes that collectively improve the internal air quality and reduce operating costs. The cells have been designed so that they can be easily upgraded for solar energy based on results in future improved technology. A portion of the northern façade is actually dedicated to ongoing research into solar cells that are to be conducted jointly by industry and RMIT. In other words, the entire building façade has the capacity to be upgraded as solar technology evolves, and may one day generate enough electricity to run the entire building.

The façade comprises a specifically detailed double-glazed inner skin in addition to an automated operable second skin shading device on each face of the building. The shading device on the second skin surrounds the entire building, from the ground floor to the roof plant level. It consists of nominally 600 mm diameter sandblasted glass disks fixed to either a horizontal or vertical aluminium axle. Each axle is fixed to the outer face of a galvanized steel cylinder of a slightly greater diameter and nominally 130 mm in depth. Twenty-one glass discs and steel cylinders are fixed together in panels of 1.8 m by 4.2 m nominal, which are supported on a galvanized secondary steel frame set out approximately 700 mm from the face of the curtain wall. These are accessed by an external service walkway on each level.

RMIT describes the building as having: "A double-skin façade with a unique external skin that incorporates glazing disks. The disks in the outer façade have the capacity to be fitted with photovoltaic collectors for harnessing solar power." It was conceived on the premise that the external discs could be rotated in two dimensions. That would have enabled the discs to track the trajectory of the sun and thereby (a) collect solar power in hot weather by deploying PV units in the discs while simultaneously shading the building from the hot sun, and (b) admit the warming sun to fall on the inside wall in cold weather. It was a fascinating idea. Now that the University has confirmed that the building lacks these capabilities, the question remains whether or not, as-built, the external wall offers effective thermal control.

On the one hand, the discs cannot be opened to admit the winter sun; on the other, the discs are made of glass that does not have reflective foil, so they cannot effectively shade the building from the hot summer sun. Moreover, existing gaps between the discs signify that 21% of the outer wall does not offer to shade. As a result, there is reasonable doubt whether the external wall qualifies as "green" in any meaningful way. RMIT has not responded to questions about the thermal efficiency and cost-effectiveness of the outer wall relative to other solutions that could have been used, nor has the University indicated if it prepared a detailed analysis of the costs and benefits of the wall before committing to the final design.

Overall, façades are an important part of the building. The future of façades is very bright, meaning that they will affect the interior and exterior of the building. They will be able to adapt to changing interior and exterior needs of the space and react to the environment. They will change the way our buildings look and impact how they function. Façades of the future will build upon the dynamic façades we see in today's market but apply a much deeper range of input and will be able to adapt to the environment more

Figure 7.7 RMIT Design Hub

Credit: Alpha from flicker: https://www.flickr.com/photos/avlxyz/6645171635

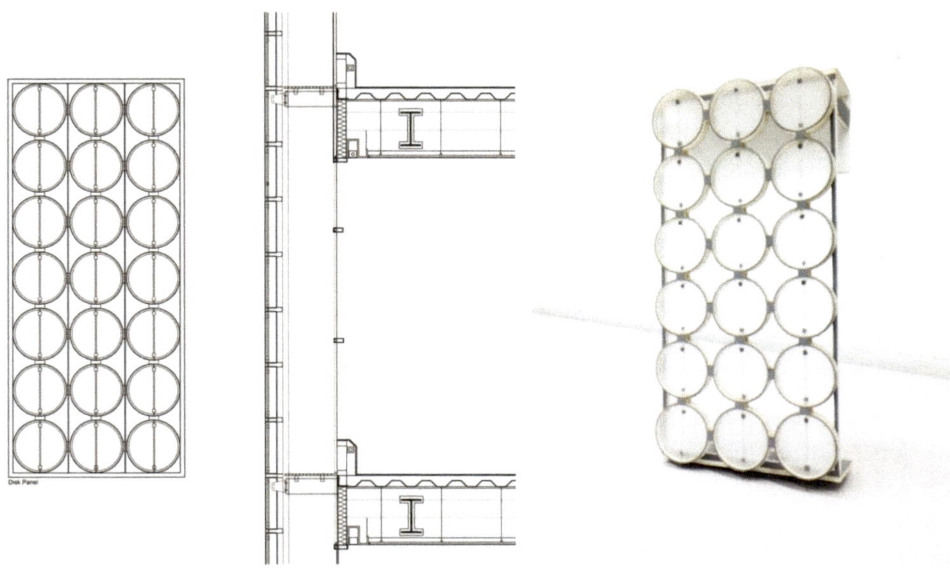

Figure 7.8 Section Drawings

closely. Façades of the future will positively influence and have a significant role in shaping the future of cityscapes.

Conclusion

In its core function, the façade protects users of a building from the outside environment. Their secondary function is to provide desirable aesthetics for a building. Given the fact that people in modern societies spend more than 90% of their time indoors, it is essential to consider building system characteristics and façade design in relation to factors of health and human comfort. However, providing energy efficiency and comfort simultaneously has been a complicated challenge for designers. As façade systems become more complex, dynamic and interactive, they will eventually respond to various external, internal and user inputs. The two significant factors that cause health and discomfort issues are derived from indoor thermal and lighting conditions that need to be regulated simultaneously in order to satisfy occupant preferences and energy objectives of a building. This challenge requires a holistic perspective toward individual preferences and building façade design. A synthesis of technology and a human-in-the-loop approach can guarantee a high degree of satisfaction and occupants' well-being.

In addition, the future interior environment introduces interactive spaces for clusters of people who have similar thermal and visual comfort needs. Therefore, the next generation of the façades will be equipped with a new control system that offers an opportunity to mediate between occupants' preferences in group settings while also optimizing energy, comfort and maximizing the use of space.

The microclimate strategy for the HVAC system provides multi-interconnected thermal microzones surrounding each occupant and can be adapted by their personal preferences, to meet energy, comfort, and health objectives. In addition, the microclimate strategy for the façade system provides dynamic visual microzones that the desired shading status can adjust based on occupant feedback and system objectives.

References

1 Iennarella, S., Serra, V., & Lo Verso, V. (2015). A novel concept of a responsive transparent façade module: Optimization of energy performance through parametric design. *Energy Procedia*, *78*.

2 Wigginton, M., & Harris, J. (2002). *Intelligent skins*. Architectural Press.

3 Gallo, P., & Romano, R. (2017). Adaptive façades, developed with innovative nanomaterials, for a sustainable architecture in the Mediterranean area, *Procedia Engineering*, *180*, 1274–1283

4 Buia, D., Nguye, T. N., et al. (2020). Enhancing building energy efficiency by adaptive façade: A computational optimization approach. *Applied Energy*, *265*, 1 May, 114797.

5 M. Ahmed, A. Abdel-Rahman, M. Bady, M., & Mahrous, E.K. (2016). The thermal performance of residential building integrated with adaptive kinetic shading system. *Int Energy J*, *16*, 97–106.

6 Hasselaar, B. (2006). Climate adaptive skins: towards the new energy-efficient façade, *WIT Trans. Ecol. Environ.*, 99 351–360, doi:10.2495/RAV060351

7 Velasco, R., Brakke, A. P., & Chavarro, D. (2015). Dynamic façades and computation: Towards an inclusive categorization of high performance kinetic façade systems. *CAAD Future Proceedings*, *527*, 563–581. doi:10.1007/978-3-662-47386-3_10

8 Haidari, H. (2011). Decisive design aspects for designing a kinetic façade: Aspects influencing the operation, figure 5. http://www.hasanahaidari.nl/images/Kinetic%20facades%20-%20HHaidari.pdf

9 Salusbury, T. (n.d.). GSW Headquarters, Berlin. https://www.arup.com/projects/gsw-headquarters

10 Ben Bacha, C., & Bourbia, F. (2016). Effect of kinetic façades on energy efficiency in office buildings – hot dry climates. *11th Conference on Advanced Building Skins*, *1*, 458–468. https://www.academia.edu/29194665/Effect_of_kinetic_façades_on_energy_efficiency_in_office_buildings_-hot_dry_climates

11 Alotaibi, F. (2015). The role of kinetic envelopes to improve energy performance in buildings. *Journal of Architectural Engineering Technology*, *4*(3), 1–5. doi:10.4172/2168-9717.1000149

Index